part of team research

positivistic tendency - tries too hard to explain everything

PIGS FOR THE ANCESTORS

Ritual in the Ecology of a New Guinea People

equilibrium model makes his thesis his conclusion - but no part of it quite pinned down -

could have argued explicitly as well as implicitly that population so near subsistence that there is no room for change

a cultural evolutionist but does not discuss it -

PIGS FOR THE ANCESTORS

Ritual in the Ecology of a New Guinea People

by Roy A. Rappaport

Yale University Press, New Haven and London

Second printing, 1970.
Designed by Marvin H. Simmons,
set in Caledonia type,
and printed in the United States of America by
The Carl Purington Rollins Printing-Office of the
Yale University Press, New Haven, Connecticut.
Distributed in Great Britain, Europe, and
Africa by Yale University Press, Ltd., London;
in Canada by McGill-Queen's University Press,
Montreal; in Mexico by Centro Interamericano de
Libros Académicos, Mexico City; in Australasia
by Australia and New Zealand Book Co., Pty.,
Ltd., Artarmon, New South Wales; in India by
UBS Publishers' Distributors Pvt., Ltd., Delhi;
in Japan by John Weatherhill, Inc., Tokyo.

Library of Congress catalog card number: 68–13926
ISBN: 0–300–01378–7

For

my parents
my wife
my children

CONTENTS

ILLUSTRATIONS

FOREWORD

Roy Rappaport's study of the functions of ritual among a primitive farming people is one of the first to be published on the ethnography of New Guinea's highland fringe. Here at the time of Rappaport's fieldwork were found what A. P. Elkin, in his 1961 review of urgent research needs, described as some of the few remaining "untouched" peoples in Australian New Guinea.[1] Further, as Elkin predicted and Rappaport now shows, cultural adaptation has taken somewhat different forms in the fringe than in the central core of the highlands. The book helps to fill major gaps in New Guinea ethnography.

Yet this is a book not just for New Guinea specialists. It should be of interest to anybody concerned with problems in functional analysis, human ecology, and the study of religion. Some remarks on certain of these problems in relation to Rappaport's study are in order here.

Functional analysis in the social sciences has been widely criticized in a number of ways. Among the most common criticisms are three that are appropriate for our consideration: (1) that functional analysis is inadequate for the explanation of the presence or origins of cultural traits and institutions; (2) that functional analysis is one-sided and almost "panglossian" in its focus on utility, harmony, integration, coherence, etc. in the status quo; and (3) that there can be no objective testing of hypotheses in functional analysis because of the lack of clear empirical import for crucial terms and concepts.

1. A. P. Elkin, 1961, Urgent research in Australian New Guinea and Papua. *Bulletin of the International Committee on Urgent Anthropological and Ethnological Research* 4, 17–25.

The first criticism is based on the recognition that more than one trait or institution may fulfill a specified function. It follows from this that demonstrating functions of particular traits cannot explain why they, rather than some functional alternatives, should be present or should have developed in a particular place or time.

The criticism is logically sound but it is also irrelevant if the object of analysis is, as for Rappaport, a demonstration of how things work rather than an explanation of why they exist or how they have come to be. In Rappaport's study, the presence of certain rituals in the cultural repertoire of the Maring people is simply accepted as given, and the problem is to show how these rituals operate in relation to various environmental processes and in relation to land use, warfare, food distribution, and other Maring activities. More precisely, the problem is to show the operation of negative feedback systems, i.e. to show how the rituals function to maintain a number of variables or activities in certain appropriate or advantageous states despite the action of disturbances tending to remove the variables or activities from those states.

Is there something "panglossian" about the choice of such a problem or is the second of the previously noted criticisms of functional analysis inapplicable? In the crude version which attributes to functional analysis Dr. Pangloss' view that all is necessarily for the best in this best of all possible worlds, the criticism certainly is inapplicable, for a concern with how extant systems operate by no means commits the analyst to the proposition that no other systems could operate better.

It nevertheless remains true that the functional analyst looks for order, coherence, equilibration, and the like rather than for their opposites, but the procedure is a defensible one on heuristic grounds. As Philip Leis has said, "The objective of revealing pattern where there seems to be only chaos or irrationality, even when pursued as an article of faith, is assuredly the most provocative and stimulating perspective to maintain."[2] And we may assume that evolutionary selection is working continually to produce pattern or order and to reduce disorder and that, accordingly, there is much order to be found, even if there is disorder too as a result, for example, of change

2. Philip E. Leis, 1965, The nonfunctional attributes of twin infanticide in the Niger delta. *Anthropological Quarterly*, 38:97–111.

in environmental conditions without the emergence of new patterns to meet them.

The preceding remarks may seem sufficient justification for a program of looking for functions, but there remains the problem of how one can know when one has found them. If our hypotheses about functions include key terms without clear empirical import, then the third of the previously noted criticisms is pertinent and the presence of hypothesized functions cannot very well be empirically validated or objectively tested for. This is so in the case of some classic statements in the literature of functionalism—for example, about the functions of various rites in reinforcing group identity or maintaining a continuity of social structure or meeting some purported needs which are never defined in any operational way. The inadequacies of such formulations have been convincingly exposed by Carl Hempel [3] and other critics.

These inadequacies are, however, not inherent in functional analysis. As Paul Collins, for example, has noted, a feasible procedure is to specify systems of operationally defined variables which are hypothesized to be maintaining some particular variables within determinable limits.[4] Part of the interest of Rappaport's study is that it constitutes a sustained attempt to isolate such systems and to validate hypotheses about their operation through the use of extensive quantitative data on the systemic variables. A way is being shown here for using empirical procedures in functional analysis. Even when Rappaport's data are insufficient for firm conclusions, the implication is not that data cannot be employed for testing the hypotheses but rather that more data need to be collected.

The fact that the systems dealt with by the indicated procedures are sets of selected variables and not the "whole" societies, cultures, or communities of traditional functional analysis is likely to invite criticism on the grounds of incompleteness. There is an easy answer to this: simply to admit that things are indeed being left out and at the same time to insist that they have to be if analysis is to proceed and to be rewarding, for the investigator can never deal meaningfully with all of the infinite number of variables confronting him and must

3. Carl Hempel, 1959. The logic of functional analysis, *Symposium on Sociological Theory*, L. Gross, ed.

4. Paul Collins, 1965, Functional analyses in the symposium "Man, culture, and animals," *Man, Culture and Animals*, Anthony Leeds and Andrew P. Vayda, eds.

therefore make a selection among them. There is, however, a further answer to be made, and this consists of noting that the analysis can become progressively comprehensive as more and more systems are dealt with and as their articulation with one another also is investigated. Biologists, in their study of feedback systems, have found that they can begin with analyses of simple components and then eventually incorporate the results of these analyses in large and comprehensive flow diagrams without impairing the validity of the original partial schemes.[5] The possibility that the same will be the case in the anthropological study of feedback systems is additional encouragement for proceeding the way that Rappaport has. As a matter of fact, he does have some things to say about the articulation of systems in which the Tsembaga local group of Maring people is involved, and the reader may find himself impressed more by how much is systematically included in Rappaport's analysis than by what is left out.

The interest of the study for human ecologists lies in large part in the data reported, i.e. data on the production and expenditure of calories, on the management of scarce protein resources, on techniques of conserving forest land, and on many other aspects of a particular nonindustrial people's adaptation to their environment. But interest for those concerned with problems in human ecology lies also in a methodological feature of the study: the inclusion of certain noncultural variables in the systems analyzed. Although there have been numerous pleas for treating cultural, environmental, and human biological variables as parts of one system, these pleas have been but little heeded by most social scientists.[6] Even among the relatively few contemporary social scientists who are especially concerned with the relation between cultural and noncultural phenomena, the prevailing tendency has been to define the cultural variables

5. H. Kalmus, 1966, Control hierarchies, *Regulation and Control in Living Systems,* H. Kalmus, ed.

6. Some recent examples of such pleas are: O. D. Duncan, 1961, From social system to ecosystem, *Sociological Inquiry,* 31:140–49; Clifford Geertz, 1963, *Agricultural Involution: The Process of Ecological Change in Indonesia,* ch. 1, Berkeley, University of California Press; D. R. Stoddart, 1965, Geography and the ecological approach: the ecosystem as a geographic principle and method, *Geography,* 50:242–51; Andrew P. Vayda, 1965, Anthropologists and ecological problems, *Man, Culture, and Animals,* Anthony Leeds and Andrew P. Vayda, eds., Washington, D.C., American Association for the Advancement of Science 78; and A. P. Vayda and R. A. Rappaport, in press, Ecology, cultural and non-cultural, *Introduction to Cultural Anthropology: Essays in the Scope and Methods of the Science of Man,* James A. Clifton, ed.

and the other ones as belonging to separate systems and then to ask about the influence of the systems upon one another.[7] In the case of many social scientists, the procedure seems to be an almost automatic one, consonant with the ingrained habit of seeing the inorganic, organic, and sociocultural as separate realms or levels of phenomena.

The alternative procedure exemplified by Rappaport's work is to make the assignment of variables to different systems depend upon either a demonstration or a testable hypothesis to the effect that the components of each system operate, at least at times, independently of the components of other systems. In other words, the discrimination of systems by this method is based on the knowledge or the expectation that the traits or variables regarded as components of any particular system do, in some way and at some times, affect one another more than they affect or are affected by extra-systemic traits or variables. The procedure has considerable heuristic value in that, as Rappaport's study illustrates, it directs attention to investigable questions about just how and when and to what degree are different traits or variables, whether cultural or noncultural, affected by one another. The same heuristic value clearly cannot obtain if it is taken as axiomatic that cultural variables belong only to strictly cultural systems.

Finally, it must be noted that Rappaport's emphasis on certain rituals as variables in the systems analyzed makes his work pertinent in the study of religion. Seeing rituals as being without practical effects on the external world has been, as Rappaport notes, a dominant view in this field of study. By elucidating the role of Maring rituals as counteracting responses to factors disturbing the people's relations with their environment, Rappaport effectively challenges this dominant view and points the way for fresh approaches in the study of religion.

Andrew P. Vayda

New York City, July, 1967

7. See, for example, Thomas G. Harding, 1960, Adaptation and stability, *Evolution and Culture,* Marshall D. Sahlins and Elman R. Service, eds., Ann Arbor, University of Michigan Press; and Marshall D. Sahlins, 1964, Culture and environment: the study of cultural ecology, *Horizons of Anthropology,* Sol Tax, ed., Chicago, Aldine.

PREFACE

The fieldwork that forms the basis of this study took place between October 1962 and December 1963 under a grant from the National Science Foundation to Columbia University. I also received personal support from the National Institutes of Health in the form of a predoctoral fellowship.

Professor Andrew P. Vayda, of the Department of Anthropology, Columbia University, was principal investigator. In addition, the expedition consisted of Professor Vayda's wife, Cherry Vayda, Allison Jablonko, also of the Columbia University Department of Anthropology, and her husband, Marek Jablonko, and my wife, Ann, and myself. Research among the Maring people has been continued under related grants by the geographers William Clarke, of the University of California, and John Street, of the University of Hawaii, by the Vaydas, who made a second visit to New Guinea in 1966, and by Georgeda Bick, of the Department of Anthropology, Columbia University. I wish to thank all of these workers both for their suggestions concerning interpretation and for the factual contributions they have made to this study.

The list of those who have offered help and encouragement before, during, and after the period of field work is long. On the way to the field E. H. Hipsley, of the Commonwealth Department of Health, Ross Robbins, of the Australian National University, and Jacques Barrau, then of the South Pacific Commission, offered invaluable advice and suggestions. John Womersley, director of the Department of Forests Herbarium at Lae identified all plant specimens, and Joseph Szent-Ivany, government entomologist, Port Moresby, took responsibility for the identification of insects. Hugh Popenoe, director of the Department of Soils, University of Florida, analyzed the soil specimens that survived the year-long trip to his laboratory. To these specialists I wish to express not only for myself, but for the others as well, our deepest appreciation.

Our life and work in the field were made easier and pleasanter through the kindness of many residents of the Trust Territory of New Guinea. R. McCormac, District Agricultural Officer, Madang, was particularly generous in providing us with personnel and equipment and in assisting us with some of our logistic problems. The personnel of both the Lutheran Mission Hospital at Yagaum and the Government Hospital at Madang acquainted us with prevalent local diseases and initiated us into such mysteries as the use of the hypodermic needle, and the Government Hospital also provided us with medical supplies.

To the staff of the Anglican Mission at Simbai we are deeply grateful not only for their assistance in overseeing our supplies and arranging for them and our mail to be sent out to us, but also for their warm hospitality on our visits to Simbai. We also wish to acknowledge with gratitude the cooperation and hospitality of Alan Johnson and Gavin Carter, the officers in charge at the Simbai Patrol Post during our stay in their area.

Several people, during visits to our field location, opened our eyes to phenomena previously unnoticed or not assigned their proper significance. The visit of Douglas Yen was a turning point in my study of horticulture, and Ralph Bulmer made tentative identifications of a large part of the avifauna.

Many members of the Department of Anthropology of Columbia University, both students and faculty, have listened to me patiently while this study was being written and have offered valuable comments. I am particularly indebted to Morton Fried, Marvin Harris, Margaret Mead, and Andrew P. Vayda.

Others who have offered valuable help and suggestions include Alexander Alland, Jacques Barrau, Elizabeth Brown, Ralph Bulmer, Marshall Childs, William Clarke, Paul Collins, Harold Conklin, Gregory Dexter, Fred Dunn, C. G. King, W. V. Macfarland, M. J. Meggitt, George Morren, Ernest Nagel, Jane Olson, David Osborne, John Sabine, Marshall Sahlins, Harold Scheffler, John Street,

Marjorie Whiting, and Aram Yengoyan. It is to Collins that I owe many of the notions concerning functional systems that are explicit or implicit in this study. Barrau, Clarke, Street, and Whiting have been particularly generous with comments, assistance, and suggestions, all of which have been appreciated although not all accepted. I personally must accept full responsibility for the shortcomings of this study.

I would like finally to express my gratitude to many Maring and Narak friends and informants.

Roy A. Rappaport

May 1967
Ann Arbor, Michigan

NOTE ON MARING ORTHOGRAPHY

Since it is assumed that the reader's primary interest is not linguistic, the orthography employed in this study is not phonemic. It is, rather, a "broad transcription" of Maring speech, and signs representing specific sounds have been selected as much for their familiarity to the English speaker as for their phonetic accuracy. The letters used in the Maring terms have English values with the following exceptions and limitations:

a as in f*a*ther
i as in m*i*ss or m*ee*t
e as in *e*rror or m*ay*
o as in b*o*ne
u as in B*u*ddha
oe as in French *oeu*f
ü as in German *ü*ber
ñ as in Spanish ma*ñ*ana
ŋ as in si*ng*
č as in *ch*urch
r is pronounced with a quick flap of the tongue tip similar to the Spanish r
initial b, d, and g have slight prenasalization (^{m}b, ^{n}d, $_{ŋ}$g)

Although the area occupied by the Maring speakers is not large, there are differences among the local groups in both the rendition of some phonemes and in some items of vocabulary. The pronunciation and terms included in this study should be understood to represent the speech of the people among whom we lived, principally the

Tsembaga, and not necessarily the speech of the entire Maring population.

It was only in the last few months of fieldwork that my command of Maring was sufficient to serve as an ethnographic tool, and in the earlier stages of fieldwork, especially, I relied upon pidgin English (hereafter referred to as "p.e."). The use of pidgin required interpreters, since no Tsembaga could speak that language. Two interpreters were used alternately, and, in matters in which ambiguity was present or error or misunderstanding suspected, the information obtained through one was checked with the other.

CHAPTER 1

Ritual, Ecology, and Systems

Many functional studies of religious behavior have had as an analytic goal the elucidation of events, processes, or relationships occurring *within* a social group of some sort. Works of Chapple and Coon (1942:507), Durkheim (1912), Gluckman (1952), Malinowski (1948), and Radcliffe-Brown (1952) may be cited as examples.

While the scope of the social unit is frequently not made explicit, it would seem that in some studies it is what Durkheim called a "church," that is, "a society whose members are united by the fact that they think in the same way in regard to the sacred world and its relations with the profane world, and by the fact that they translate these common ideas into common practices" (Durkheim, 1961: 59). Frequently, however, it is a smaller and more bounded group that provides the context within which the role of ritual in relation to other aspects of culture is studied. Such units, composed of aggregates of individuals who regard their collective well-being to be dependent upon a common body of ritual performances, might be called "congregations." In many small-scale societies the congregation is coterminous with the local group. Such is the case in this study of the Tsembaga, a group of shifting horticulturalists living in the Bismarck Mountains of New Guinea.

This study differs from those mentioned above, however, in that its main concern is not with the part ritual plays in relationships occurring within a congregation. It is concerned, rather, with how ritual affects relationships between a congregation and entities *external* to it.

Maring rituals are conventionalized acts directed toward the involvement of nonempirical (supernatural) agencies in human affairs.[1] Although suggestions have been made by a number of writers concerning the possible role of ritual in the adjustment of social groups to their environments,[2] this problem has not engaged the attention of many students of either religion or human ecology. Some writers, interpreting what appear to be economically wasteful practices, have taken the view that ritual frequently interferes with the efficient exploitation of the environment.[3] Others have tacitly assumed that the empirical relations of the congregation to its environment remain unaffected by ritual performances.

Indeed, one of the important functionalist theories of religion is based upon such an assumption of empirical independence between ritual and the world external to the congregation. It asserts that since men are unable to control many of the events and processes in their environments that are of crucial importance to them, they experience a feeling of helplessness. This helplessness produces anxiety, fear, and insecurity. The performance of rituals suppresses anxiety, dispels fear, and provides a sense of security. A statement by George Homans summarizes this line of thought nicely:

> Ritual actions do not produce a practical result on the external world—that is one of the reasons why we call them ritual. But to make this statement is not to say that ritual has no function. Its function is not related to the world external to the society, but to the internal constitution of the society. It gives the members confidence, it dispels their anxieties, it disciplines their social organization. [1941:172]

No arguments will be raised here against the psychological or sociological functions that Homans and others have imputed to ritual.

1. This statement is merely descriptive. It does not purport to be a definition of Maring ritual, much less ritual in general.

2. Among those who have made such suggestions are Brown and Brookfield (1958, 1963), Cook (1946), Firth (1929, 1950), Freeman (1955), Izikowitz (1951), Moore (1957), Stott (1962), and Vayda, Leeds, and Smith (1961). Concern here is with ritual, rather than taboo, which may be described as the supernaturally supported proscription of physically feasible behavior. It may be mentioned, however, that a number of writers have noted the possible role of taboos in the conservation of resources, and a recent paper by Harris (1965) has elucidated the critical role of the taboo against the consumption of beef in the human ecology of India.

3. See, for example, Luzbetak (1954:113) concerning waste at a New Guinea pig festival.

But it will be argued that in some instances ritual actions *do* produce a "practical result on the external world." In some instances the "function" of ritual *is* related to the "world external to the society." Among the Tsembaga, ritual not only expresses symbolically the relationships of a congregation to components of its environment but also enters into these relationships in empirically measurable ways.

To state simply that a people's ritual actions may measurably affect components of their environment is of course to state the obvious, if not the trivial. If, to perform a ritual, a man cuts down a tree, the environment is affected. If ritual requires that large numbers of visitors be entertained at prolonged and lavish feasts, extra-large gardens will be planted, or more wild animals than usual will be killed, or particularly large quantities of wild vegetables will be collected. All of these are actions that by any definition of environment affect the environment.

But it is not the obvious effects of isolated ritual requirements that are the central concern here. The interest, rather, is in the ways in which ritual mediates critical relationships between a congregation and entities external to it. Among the Tsembaga, and other Maring-speaking groups in New Guinea, through ritual the following are effected:

1. Relationships between people, pigs, and gardens are regulated. This regulation operates directly to protect people from the possible parasitism and competition of their pigs and indirectly to protect the environment by helping to maintain extensive areas in virgin forest and assuring adequate cultivation-fallow ratios in secondary forest.
2. The slaughter, distribution, and consumption of pig is regulated and enhances the value of pork in the diet.
3. The consumption of nondomesticated animals is regulated in a way that tends to enhance their value to the population as a whole.
4. The marsupial fauna may be conserved.
5. The redispersal of people over land and the redistribution of land among territorial groups is accomplished.
6. The frequency of warfare is regulated.

7. The severity of intergroup fighting is mitigated.
8. The exchange of goods and personnel between local groups is facilitated.

Ritual will be regarded here as a mechanism, or set of mechanisms, that regulates some of the relationships of the Tsembaga with components of their environment. The terms *regulate* or *regulation* imply a system; a system is any set of specified variables in which a change in the value of one of the variables will result in a change in the value of at least one other variable. A regulating mechanism is one that maintains the values of one or more variables within a range or ranges that permit the continued existence of the system.

It should be emphasized that neither the Tsembaga nor any of the actual components of their environment are themselves variables. As Hagen (1962:506) has put it, "A variable is a single dimension of an entity, not the entity itself." The size of the Tsembaga population, for instance, but not the Tsembaga themselves might be regarded as a variable. The amount of land in cultivation, expressed in acres, might be designated a variable, but the gardens themselves could not be.

The systemic relationships described in this study are more than regulated; they are self-regulated. The term *self-regulation* may be applied to systems in which a change in the value of a variable itself initiates a process that either limits further change or returns the value to a former level. This process, sometimes referred to as "negative feedback," may involve special mechanisms that change the values of some variables in response to changes in the values of others. Thermostats, for instance, may be regarded as mechanical regulating mechanisms in systems in which measurable quantities of heat emanating intermittently from a controlled source and the temperature of a surrounding medium are variables. It will be argued here that Tsembaga ritual, particularly in the context of a ritual cycle, operates as a regulating mechanism in a system, or set of interlocking systems, in which such variables as the area of available land, necessary lengths of fallow periods, size and composition of both human and pig populations, trophic requirements of pigs and people, energy expended in various activities, and the frequency of misfortunes are included. There are numerous additional variables to be considered as well. While it has not been possible in all cases, numerical values have been assigned to most of the variables on the basis of measurements performed in the field.

As Collins (1965:281) has pointed out, "Functional analysis makes no prescription concerning the nature of the variables . . . constituting the system." The selection of variables is a product of hypotheses concerning possible interrelations among the phenomena under investigation, and these, in turn, flow from the interests and theoretical conceptions of the analyst. The hypotheses that have led to the selection of variables in this study have already been stated in the form of a number of propositions concerning the role of ritual in the adjustment of the Tsembaga to their environment. Underlying these hypotheses is the belief that much is to be gained by regarding culture, in some of its aspects, as part of the means by which animals of the human species maintain themselves in their environments. There should be no conceptual difficulty in treating culture much as one would the behavior of other animals. As the sociologist Hawley pointed out over twenty years ago:

> Culture is nothing more than a way of referring to the prevailing techniques by which a [human] population maintains itself in its habitat. The component parts of human culture are therefore identical in principle with the appetency of the bee for honey, the nest-building activities of birds, and the hunting habits of carnivora. To argue that the latter are instinctive while the former are not is to beg the question. Ecology is concerned less now with how habits are acquired, than with the functions they serve and the relationships they involve. [1944:404]

Hawley's statement has important methodological and theoretical implications for cultural anthropology, and Vayda and I (Vayda and Rappaport, in press) have noted some of these as follows:

> Consistent with usage in [general] ecology, the focus of anthropologists engaged in ecological studies can be upon human populations and upon ecosystems and biotic communities in which human populations are included. To have units fitting into the ecologists' frame of reference is a procedure with clear advantages. Human populations as units are commensurable with the other units with which

> they interact to form food webs, biotic communities and ecosystems. Their capture of energy from and exchanges of material with these other units can be measured and then described in quantitative terms. No such advantage of commensurability obtains if cultures are made the units, for cultures, unlike human populations, are not fed upon by predators, limited by food supplies, or debilitated by disease.

The adoption of populations and ecosystems as units of analysis, it should be stressed, does not require any sacrifice of anthropology's primary goal of elucidating cultural phenomena. Quite to the contrary, such a procedure can make important contributions to the attainment of that goal. A population may be defined as an aggregate of organisms that have in common certain distinctive means for maintaining a set of material relations with the other components of the ecosystem in which they are included. The cultures of human populations, like the behavior characteristic of populations of other species, can be regarded, in some of their aspects, at least, as part of the "distinctive means" employed by the populations in their struggles for survival. It has been suggested by the biologist G. G. Simpson (1962:106) that the study of cultural phenomena within such a general ecological framework may provide additional insights into culture, "for instance, in its adaptive aspects and consequent interaction with natural selection."

The lead suggested by Hawley and Simpson has been followed in this study. The Tsembaga are regarded here as an ecological population in an ecosystem that also includes the other living organisms and the nonliving substances found within the boundaries of Tsembaga territory. The rituals upon which this study focuses are interpreted as part of the distinctive means by which a population, the Tsembaga, relates to the other components of its ecosystem and to other local human populations that occupy areas outside the boundaries of Tsembaga territory.

It should be made explicit perhaps that the operation of ritual as a regulating mechanism is not necessarily understood by the Tsembaga. In the language of sociology, regulation is a "latent function" (Merton, 1949:19ff) of Tsembaga ritual. The Tsembaga themselves see the purposes of the rituals as having to do, rather, with the relations of people to various spirits—for the most part, those of deceased ancestors. The rituals with which this study is concerned are conceived by the Tsembaga to maintain or transmute their relationships with these nonempirical entities. It would be possible to elucidate the regulatory functions of Tsembaga ritual without reference to Tsembaga conceptions, but it is reasonable to regard the conceptions of a people as part of the mechanism that induces their behavior. Native views of the rituals will therefore be included in this study. The inclusion of native views in a description of ecological relations, moreover, permits us to ask some important questions concerning ideology. We may, for example, ask whether or not actions undertaken in reference to understandings that are not empirically valid are, nevertheless, appropriate to the actual situation in which the actors find themselves.

CHAPTER 2

The Tsembaga

THE PEOPLE, THEIR LOCATION, AND THEIR LINGUISTIC RELATIONSHIPS

The Tsembaga are a group of about 200 Maring-speaking people living in the Madang District of the Australian-administered Territory of New Guinea. The name *Tsembaga* as a designation for this cluster of clans is recent, having been first used in this way by a government patrol officer in 1960. Previously Tsembaga had denominated only one of the constituent clans; the cluster as a whole was unnamed. Since 1960, however, the use of Tsembaga to designate the entire local group has not only become enshrined in the census book and on maps, but has been taken up by the people themselves.

The Tsembaga occupy an area of slightly more than three square miles at approximately 5° south latitude, 145° east longitude on the south wall of the Simbai Valley, which separates, in this region, the Bismarck Range to the south and a spur of the Schrader Range to the north. The land is very mountainous, rising within the confines of the small Tsembaga territory from 2,200′ at the river to 7,200′ at the top of the Bismarck ridge, and is, for the most part, heavily forested. Most of the area above 5,000′ shows no signs of ever having been cut, while at lower altitudes, secondary forest prevails, although some small remnants of primary forest and a few patches of grassland are to be found.

The amount of contact with Europeans to which the Tsembaga had been exposed at the time of fieldwork (October 1962 to December 1963) was slight. The first patrol of the Australian government to enter the area crossed their territory in 1954, but a second did not appear for several years. The Simbai Valley was pacified in 1958,[1] but the government did not officially regard the area as controlled until 1962. A *luluai* (headman) and a *tultul* (assistant headman) were appointed by the Australian government in 1959, and a second *tultul* was appointed in 1961. However, in 1963 the prerogatives of these government headmen generally remained restricted to dealings with the government. Both those who held the titles and the people at large regarded these offices as largely irrelevant to internal affairs.

At the time of fieldwork only two young Tsembaga men had gone out of the Maring area as indentured laborers, and they had not yet returned.[2] Although the Anglican mission had maintained stations with two other Simbai Valley groups for some years, the Tsembaga remained unmissionized and their religious beliefs and practices included no apparent European elements. No Tsembaga spoke pidgin English, and only a few of the younger men had ever visited the government patrol post, established in 1959 at the headwaters of the Simbai River about 25 miles away by trail, until our arrival.

Indirect contact began, of course, much earlier. The first steel tools came to the Tsembaga in the late 1940s, and by the early 1950s stone implements had been completely replaced. Epidemics, probably of European origin, also entered the area long before white men. A dysentery epidemic, bearing the pidgin name *sikman*, killed a large number of people in the early 1940s, and there is evidence of measles several years later.

Cargo cult also came to the Tsembaga long before white men. In the early 1940s, news of impending flood, earthquake, and ancestral return came to them from the north, along with accounts of unfamiliar treasures that they might receive. Although they followed instructions, building special houses and so on, the ancestors did not return bearing gifts amidst flood and earthquake, and the disillusioned Tsembaga have remained more or less impervious to cargo cult talk ever since.

In sum, while the arrival of Europeans and European commodities have produced important changes, particularly through the in-

1. Since our departure from the field in December 1963 some small-scale, short-lived armed altercations are reported to have occurred among Simbai Valley Maring.

2. Fifteen young Tsembaga men are reported to have been recruited on two-year labor contracts early in 1964.

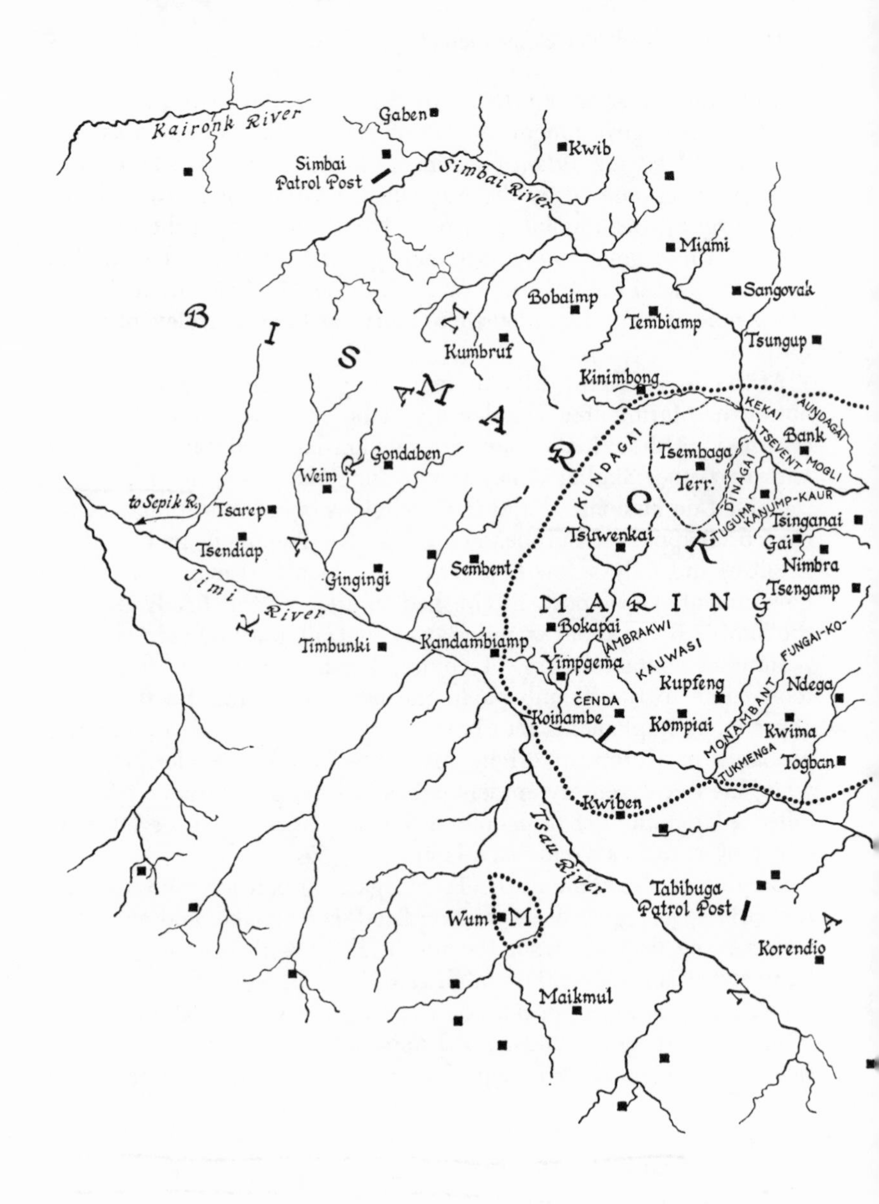

Kaironk River
Gaben
Simbai Patrol Post
Kwib
Simbai River
Miami
Sangovak
Bobaimp
Tembiamp
Tsungup
Kumbruf
Kinimbong
KEKAI
AUNDAGAI
Bank
TSEVENT
MOGLI
Tsembaga Terr.
KUNDAGAI
DINAGAI
Gondaben
Weim
to Sepik R.
Tsarep
Tsendiap
B I S M A R C K
Tsuwenkai
TUGUMA
KANUMP-KAUR
Tsinganai
Gai
Nimbra
Gingingi
Sembent
M A R I N G
Tsengamp
Jimi River
Bokapai
AMBRAKWI
KAUWASI
FUNGAI-KO-
Timbunki
Kandambiamp
Yimpgema
Kupfeng
Ndega
ČENDA
Koinambe
Kompiai
Kwima
MONAMBANT
TUKMENGA
Togban
Kwiben
Tsau River
Tabibuga Patrol Post
Wum
M
Korendio
Maikmul

Location of the Maring and Their Neighbors

troduction of steel tools and the suppression of warfare, the degree to which the adaptation of the Tsembaga to their immediate environment has been affected is limited. Steel tools and a few new crops, such as maize and certain varieties of sweet potatoes, have without doubt affected agricultural production, but the Tsembaga remain subsistence gardeners, making gardens in accordance with patterns that prevailed when there was no maize and when they had only stone tools with which to work.

The Tsembaga are one of more than twenty similar local groups of Maring-speaking people occupying territories in the middle Simbai and Jimi Valleys. The size of these groups ranges from a little over 100 to 900 persons, and in all there are about 7,000 Maring (or, as the language is occasionally and somewhat mysteriously labeled on linguistic maps, Yoadabe-Watoare) speakers. Wurm (1964:79) classifies Maring as a member of the Jimi subfamily of the Central Family of the East New Guinea Highlands Stock, which includes most of the languages of the East New Guinea Highlands (Micro) Phylum. The Maring area is the most northerly of those occupied by speakers of Central Family languages; the land to the north and west of the Maring is held by speakers of Karam and Gants, languages only distantly related to the East New Guinea Highland Stock.

The Central Family, which includes at least fourteen languages with a total of 286,000 speakers (Wurm, 1964:79), occupies a large area in the Eastern, Western, and Southern Highlands districts. This geographical distribution suggests that the ancestors of the Tsembaga and other Maring entered the general area that they presently occupy from the south. The presence of extensive tracts of unoccupied virgin forest to the north and east of the easternmost Maring groups supports this view. Tsembaga tradition, moreover, places the origins of four of its five constituent clans in the Jimi Valley in the third or fourth generation ascending from men who were middle-aged in 1962 and 1963. Tsembaga tradition is unreliable as history, but recent arrival in their present location is also suggested by such negative evidence as the lack of any clear signs that the environment has been degraded, and such positive evidence as the large size of the trees composing the secondary forest in which gardens are cut, and the reports of many older informants, who state that extensive tracts of primary forest existed in the lower altitudes until forty or fifty years prior to fieldwork.

Details of Tsembaga subsistence procedures will be discussed in the next chapter, and in later chapters the relationship of ritual to subsistence practices and other aspects of the Tsembaga adaptation will be analyzed. It may be helpful to mention here, however, that the Tsembaga and other Maring are bush-fallowing, or swidden, horticulturalists, planting their gardens in secondary forest. Their starch staples include taro, yam, sweet potato, manioc, and bananas, but they also enjoy a large number of other crops, including sugar cane, many greens, and the fruit of the *Pandanus conoideus* (p.e.: *marita*). Pig husbandry is also important. Hunting, trapping, and gathering also play a part in Tsembaga subsistence; feral pig, cassowary, and birds are the most important quarry taken in hunting, and marsupials are the animals most frequently caught in traps. The yields of gathering include not only some vegetable foods, notably edible ferns, but also timber, animal and vegetable fibers used for clothing, vines used as rope, and leaves, fruits, and earths used as dyes.

The technology of the Tsembaga is simple. Only the digging stick, the steel ax, and the bushknife are used in gardening. Bows and arrows are used in hunting, and also form, along with spears, axes, and wooden shields, the technology of warfare. Traps include snares, deadfalls, and pits. Gourds and bamboo tubes are used as containers, and bamboo tubes occasionally serve as cooking vessels also. Aside from these there are no others; most food is prepared directly on the fire or in an earth oven. Net bags, loin cloths, caps, and string aprons are woven from a variety of fibers; intricately woven waist and arm bands are made from orchid stems, and some bark cloth is made. Prior to the establishment of the patrol posts in the Jimi and Simbai Valleys in 1956 and 1959 respectively the Tsembaga manufactured salt by boiling water obtained from mineral springs. Much of their salt production in earlier times was traded south over the mountain in exchange for stone ax blades, which were quarried, shaped, and polished in the Jimi Valley.

Men and women live in separate houses. During 1962 and 1963 the men's houses, which accommodate males above the age of seven or eight, ranged in number of occupants from two to fourteen. Each married woman and widow has a separate house, where she resides with her unmarried daughters, small sons, and pigs. Most cooking takes place at the women's houses. Although the interior arrangements of men's and women's houses are somewhat different, their size and construction is similar. Both are framed in light timber, with roofs and walls thatched with *pandanus* leaves. The dimensions of men's houses range from 7'x20' to 10'x35', with a height at the ridge pole of 4½' to 6'. Women's houses are of comparable length and breadth, but on the average they are lower than the men's. A portion of the interior of a woman's house, however, is devoted to the shelter of pigs, each of which is accommodated in a separate stall with its own entrance from the outside.

DEMOGRAPHY AND PHYSIQUE

Both the subsistence procedures and trophic requirements of the Tsembaga will be assessed in the next chapter and an attempt will be made to estimate the maximum number of people who may pursue their subsistence activities on Tsembaga territory. Among the biological characteristics of the Tsembaga population that are germane to these and other assessments and estimates to be made later are: (1) the total number of individuals included in the population, (2) the age and sex composition of the population, and (3) the average body sizes of the constituent individuals. Information concerning Tsembaga demography and ontogeny can only be outlined here. Discussions of underlying factors will be published elsewhere by other workers.

Population Size

Between October 1962 and December 1963, the local population of the Tsembaga territory varied from 196 to 204 persons (the larger figure will be used in later computations).

There is considerable evidence to indicate that the population had been considerably larger before the period of fieldwork. For one thing, about twenty-five Tsembaga were living sororilocally, uxorilocally, or matrilocally with other local groups in 1963. These people, or their parents, had lived on Tsembaga territory until military defeat in 1953 forced the Tsembaga to flee. Most Tsembaga had returned to their own territory by 1963, but these twenty-five still remained elsewhere. Informants agree, further, that in earlier times there were many more people, and genealogies bear them out. They blame disease for much of the population decrease, although they admit that many were also killed in fighting. The early years of contact are especially dangerous to groups such as the Tsembaga, for they are exposed to new causes of death, particularly disease, before the old ones, especially warfare, are suppressed.

Considering the number of Tsembaga presently residing elsewhere, and considering the consensus among informants and the support given to such consensus by genealogies, a population size of 250 to 300 persons in the 1920s and 30s is not unlikely.

Population Structure

The age-sex composition of the Tsembaga population is presented in Table 1. Some of the characteristics of Tsembaga population structure, particularly the imbalance of males and females in the younger categories, and the change in the male-female ratio at estimated age twenty-five, require comment.

The Tsembaga, who are happy to admit that they kill one or both of twins, unanimously deny practicing female infanticide. They themselves point to the economic importance of women and suggest that the killing of female infants would be foolish. The reversal of the usual ratio in the estimated five- to ten-year-old category lends support to their protestations.

It is not possible to arrive at any explanation here. Data are insufficient to say with any certainty whether the disparity is a result of differential frequencies in male and female births, or differences in the rate of survival of males and females, or both. Data are sufficient to indicate, however, that a deficiency of females, although especially marked among the Tsembaga, is not peculiar to them. A. P. Vayda's census figures indicate totals of 3,722 males and 3,420 females among the Maring and only one local population where females outnumber males. Nor is such a disparity peculiar to the Maring. An excess of males exists among the people of the Tor District in West Irian. As is the case among the Tsembaga, the imbalance between the sexes is most marked in the younger categories, and Oosterwal (1961:37f) attributes it to a greater frequency of male births.

Table 1. The Structure of Tsembaga Population, November 1963 (total 204)

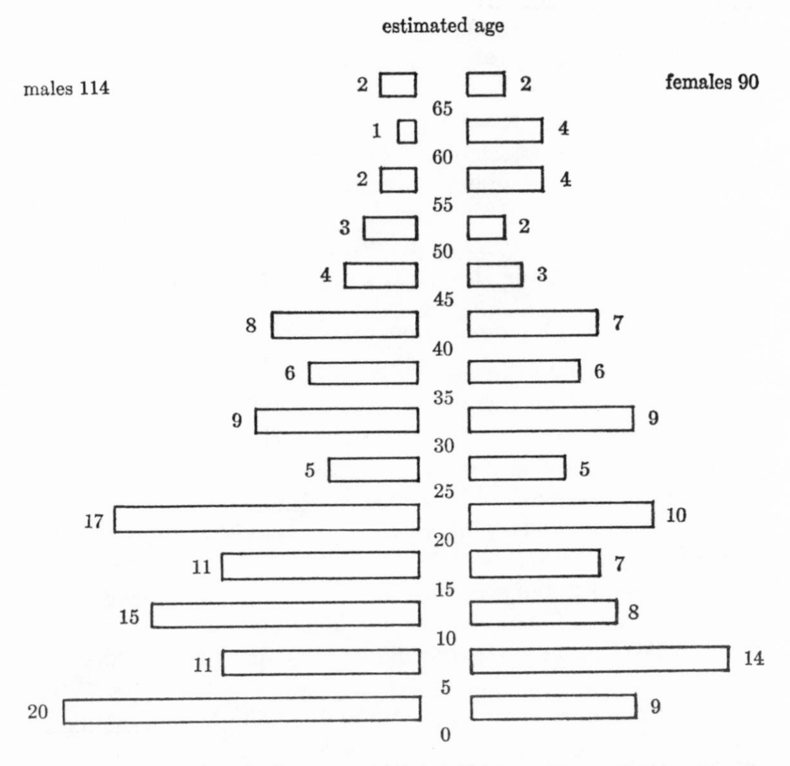

It must be noted that the sex imbalance exists despite the fact that among the Tsembaga, at least, males suffered heavier losses in warfare than females. The change in the Tsembaga male-female ratio at estimated age twenty-five is quite clearly an effect of differences in casualty rates.

Body Size

The Tsembaga are very small in stature. The average height of forty-nine adult females was 54.3″, the range being 51.75″ to 57.75″. Their average weight was 84.4 pounds, with a range of 75 to 100 pounds. The average weight of fifty-nine adult males was 102.6 pounds. Their weights ranged from 87 to 140 pounds.

Since many young men wear their hair in massive coiffures hardened with grease extracted from the fruit of the *Marita pandanus,* I could only guess at the location of the top of the skull. The average of the heights recorded for adult males, 58.1″, must therefore be regarded as an approximation. The extremes, however, 53.25″ and 62.5″, are represented by men whose hair was cut short. To what extent Tsembaga stature is an expression of genetic constitution, and to what extent it is a function of dietary deficiency, parasite infestation, and other pathology remains to be determined through future research.

POPULATION DISPERSION AND SOCIAL ORGANIZATION

Although a quantity of a necessary resource, such as arable land, sufficient to meet the trophic requirements of all the Tsembaga exists within the borders of their territory, this does not necessarily indicate that each Tsembaga has enough to meet his needs. Many societies are characterized by differentials in access to resources between individuals or between groups. In some societies this is a concomitant of social stratification. In others it may be the result of differences in the demographic fortunes of land-holding descent groups. It is therefore necessary to describe the social means by which the Tsembaga are dispersed over their territory in subsistence activities.

The Tsembaga form a single territorial unit as far as defense is concerned, and all Tsembaga may hunt, trap, and gather in any part of the territory. These rights in nondomesticated resources are exclusive. That is, members of other local groups do not enjoy rights in nondomesticated resources to be found on Tsembaga territory, and, conversely, the Tsembaga do not have such rights on the territories of other groups.

Although the entire territory is open to the hunting and gathering activities of all Tsembaga, it is divided into "subterritories," smaller areas claimed by less inclusive groups. Membership in these smaller groups is particularly important in reference to rights in garden land.

The Tsembaga are organized into five putatively patrilineal clans (*kai:* root; or *yu kai:* root of men): Merkai, Tomegai, Tsembaga, Kwibigai, and Kamuŋgagai, which ranged in resident membership from fifteen to seventy-eight persons, including in-married wives in 1963. These clans are the most inclusive units claiming a common ancestry, but such ancestry can be demonstrated only in the case of the small Kwibigai clan (twenty-one members), all of whose members are descended from an immigrant who arrived among the Tsembaga in the late nineteenth century.

The two smallest clans, Tomegai (fifteen members) and Kwibigai, are unsegmented, but the three largest clans, Kamuŋgagai, Tsembaga, and Merkai, are each divided into three smaller segments, or subclans, also termed *kai* or *yu kai*, ranging in resident membership from seven to thirty-seven persons. Subclan men claim descent from an ancestor less remote than the common clan ancestor, but this descent cannot be demonstrated in all cases. Subclans are also named, and in the cases of all three segmented clans the subclans bore the same names: *Wendekai, Amangai,* and *Atigai* (*Wend:* oldest; *Amang:* middle; *Ati:* youngest).

Ideally, it is a clan that should claim a subterritory, for the Tsembaga, like other Maring, associate territory with agnation. But although Tsembaga is composed of five clans, the subterritorial division of Tsembaga territory is only tripartite. The main portion of the territory is divided into three adjacent strips running from ridge top to the Simbai River, with the Kamuŋgagai claiming the western strip, the Merkai the eastern, and the Tsembaga, Kwibigai, and Tomegai clans jointly claiming the central subterritory.

Although marriage rules stipulate clan exogamy, exogamy in fact seems to correlate more closely with subterritoriality. The Tsembaga, Tomegai, and Kwibigai clans, which share a common subterritory, are antigamous in reference to each other, and in the past marriages between members of separate subclans within the Merkai clan were contracted. I believe that these intraclan marriages took place in conjunction with a process of clan fission and perhaps subterritorial separation, a process that was subsequently reversed (as will be discussed later).

While the three subterritorial groups seem to be exogamous, the Tsembaga as a whole are not. Indeed, there is an explicit and statistical preference for marriage to women of proximate origin. Of the fifty wives and widows residing on Tsembaga territory in 1963, 44% were Tsembaga in origin, and an additional 22% came from the Tuguma immediately to the east. The remaining 34% came from nine other local groups, in most cases occupying territories across the river or over the mountain.

Subterritoriality is expressed ritually. Claims to subterritories are ratified on certain occasions by planting on them small trees or bushes called *"yu miñ rumbim"* (*yu min:* men's "souls"; *rumbim: Cordyline fruticosa* (L.), A. Chev.; *Cordyline terminalis,* Kunth; p.e.: *tanket*). Every adult male member of the subterritorial group participates in this ritual by grasping the *rumbim* as it is planted, thus symbolizing both his connection to the land and his membership in the group that claims the land.

The rituals surrounding the *rumbim* provide an additional criterion for distinguishing the Tsembaga from adjacent groups, for the planting, and later the uprooting of *rumbim* by the three Tsembaga subterritorial groups, and these three groups only, are synchronized, usually taking place on the same day. Moreover, all Tsembaga join together for subsequent rituals. It is on the basis of their coordination of some of these rituals and their joint and exclusive participation in others that we may distinguish the Tsembaga as a single congregation distinct from all others.

Subterritories are subdivided into smaller areas claimed by smaller putatively patrilineal units, either clans or subclans. Each of these groups holds a number of noncontiguous tracts scattered throughout the full altitudinal range of the subterritory. In the arable zone these tracts are further subdivided into garden sites, generally less than an acre, claimed by individual men.

Males have rights in the entire estate in garden land of the subterritorial group by virtue of clan and subclan membership. Although individual men claim title to particular garden sites either through patrilineal inheritance or by clearing virgin forest, adjustments of inequities are readily made. If a man is short of inherited sites, he simply asks a better-endowed member of his subclan for land, and a transfer in perpetuity seems always to be made. An individual's title to garden sites thus amounts to no more than stewardship for his subclan. Similarly, the title of a subclan in a tract of land may be regarded as stewardship for the entire clan. If a subclan is short of land, not only may its individual members ask for and receive grants in land from members of better-endowed subclans, but entire tracts may be granted it by other subclans.

In the case of the subterritorial cluster of three clans, adjustments of inequities in land holdings are also readily made between members of the three separate clans. That is, as far as access to land is concerned, these three clans relate to each other as do the subclans of a single clan. These three clans, at present antigamous in reference to each other, are, or believe themselves to be, related by connections through females in ascending generations when intermarriage among them is either known or believed by them to have taken place. That is, the members of the three separate clans regard each other as cognates, even though actual connections frequently cannot be traced. It could be argued that although this multiclan structure remains de jure a cluster of three agnatic descent groups, since the principle of filiation is patrilineal connection to one of the three constituent clans, it may be regarded as, de facto, a corporate cognatic descent group since its combined membership has rights in a common estate in land. I shall therefore refer to it as a "cognatic cluster." Although there is no term in the Maring language to designate the class of structures I would label "cognatic clusters," it may be noted that the existence of particular cognatic clusters as distinct units are recognized nominally. Thus, the cluster that includes the Kwibigai, Tomegai, and Tsembaga clans is known as "Tsembaga-Tomegai."

Regardless, however, of whether rights to land are based upon agnation or cognation, within the three subterritorial groups serious differences in land holding are ameliorated as quickly as they become apparent. It may be suggested that the ease of transfer displayed by Tsembaga land tenure practices is particularly advantageous among societies organized into small groups, for small groups are highly vulnerable to sudden and drastic demographic fluctuations independent of those experienced by their neighbors. Not only do the transfers avoid social and economic inequities, but they protect the environment from overexploitation in some areas while other areas remain underexploited.

But the subterritorial groups are themselves small and thus highly subject to independent demographic fluctuations. Grants in land are, however, frequently made by members of one subterritorial group to members of another. In a sample of 381 gardens made in 1961, 1962, and 1963, the gardener worked the land of a subterritorial group of which he was not a member in 94, or 24.7 percent, of the cases. Thus, the use of land for gardening is not exclusive to the agnatic or cognatic group that claims the land. To put it in the converse, a man frequently makes gardens on lands to which his agnatic or cognatic land-holding group has no claim. While titles to three distinct areas are recognized by the Tsembaga to be vested in three discrete groups, Tsembaga gardens are very much intermingled.

It should perhaps be made clear that the pattern of residence is largely independent of land use. Distances are not great, and a man may, and usually does, maintain patrilocal residence while making gardens on land received from other subterritorial groups within Tsembaga. The actual residence pattern may be called "pulsating." At a certain point in the ritual cycle there is a high degree of nucleation, when all or almost all houses are located in the general vicinity of a dance ground. However, Tsembaga subsistence procedures, particularly those concerned with pig husbandry, militate against nucleation, and small clusters of houses and scattered homesteads are the rule at most times.

Because of the easy access of all males to garden land all over the territory, it is reasonable to assume that population density figures, which will be presented in the next chapter, reflect a condition existing throughout Tsembaga territory and not averages of what might be a number of very different densities in different areas. In other words, for purposes of estimating intensity of exploitation an even distribution of the Tsembaga over the available land may be assumed.

That all of the Tsembaga, and the Tsembaga alone, enjoy common and almost exclusive access to the resources of a jointly defended territory has a further theoretical and methodological implication. They, separate from neighboring groups, constitute a unit in a set of material exchanges with the populations of other species that also inhabit their territory. They form a population in the ecological sense of the term, and I shall refer to them and similar units as "local populations."[3]

There are, however, differences in the nature and pattern of land transfers between members of constituent units of the same subterritorial group on the one hand and between members of separate

3. My terminology is somewhat different from that employed by Vayda and Cook (1964). *Local population* in my usage corresponds to *clan cluster* in theirs.

subterritorial groups on the other. In contrast to transfers between members of separate constituent units of the same subterritorial group, which were in 28, or 85%, of the cases in the sample grants in perpetuity, only 35% of the transfers between members of separate subterritorial groups were grants in perpetuity. The remainder were usufructory grants made by men to cognates or to wives' agnates, in most cases. Of the 33 grants in perpetuity made by members of one subterritorial group to members of another, 26, or 79%, were made by men to *husbands of female agnates*. Such affinal grants first take place when it becomes apparent that a marriage will be an enduring one, but in most cases additional grants are made throughout the lives of the principals. These grants, which confer rights *only* in specified sites and *not* in the entire estate of the grantor, do not require the grantee to take up uxorilocal residence, and he seldom does. In most cases sites received from a wife's agnates only supplement the lands to which a man has rights by virtue of patrifiliation, lands which he usually continues to rely upon most heavily. Whether or not he needs lands received through affinal grants, however, he is likely to use them, for he is under pressure from his wife and his affines to do so. His wife will likely wish to make some of her gardens on the land of her agnates so that she may visit them frequently. On the other hand, a woman's agnates encourage her husband to take up these lands for less sentimental reasons. From their point of view, indeed, the affinal grant provides a means for retaining access to a portion of the labor of female agnates after their marriages. The sites transferred to a woman's husband are usually adjacent to those being cultivated by her agnates; thus, a woman can without great inconvenience assist an unmarried brother or a widowed father with his garden, and she often does.

In light of the considerations involved in cultivating land received from wives' agnates, it is not surprising that these sites, although said to be transferred in perpetuity, often revert to the grantors after one generation. The pressures that induce a man to cultivate land received from his wife's natal group are absent from his relationship with his mother's natal group. If, however, the lands his father has received through affinal transfers are convenient to a man's residence he is likely to continue to use them. Therefore, affinal transfers between adjacent subterritorial groups are likely to endure through several generations. The result is that the garden sites claimed by members of adjacent subterritorial groups become intermingled. Such intermingling has proceeded far among the Tsembaga, particularly between the cognatic cluster, which occupies the central subterritory, and the Kamuŋgagai clan, which occupies the subterritory to its west. Indeed, the two subterritories seem to be fusing. People say that the land of the two groups is or is becoming one, and many younger men claim to be ignorant of the location of the boundary between them. Older men are in agreement concerning its location but invariably state that this boundary no longer means as much as it did in former times.

It may be noted that the closeness of the relationship between the Kamuŋgagai clan and the Tsembaga-Tomegai cognatic cluster is nominally recognized: the two subterritorial groups are frequently referred to in aggregate as "Kamuŋgagai-Tsembaga." Although there is no corresponding term in the Maring language, I would designate as "affinal clusters" structures such as Kamuŋgagai-Tsembaga, which consist of two adjacent subterritorial groups whose lands are becoming intermingled through affinal transfers.

In connection with the gradual obliteration of the boundary between Tsembaga-Tomegai and Kamuŋgagai it is interesting to note that four of the new transfers in perpetuity of land between them included in the sample were not made on the basis of current affinal connections, but on the basis of affinal connections in ascending generations, that is, cognatic connections. The number is small, but I believe that it both marks a trend and illuminates the processes of group formation. Evidence is insufficient, but it may be suggested that cognatic clusters and affinal clusters represent two points in a continuous process of land amalgamation. Through intermarriage, and then by grants in perpetuity through affinal connections, the land of adjacent groups becomes intermingled. Since neither a man nor his subclan "brothers" should take a wife from his mother's natal clan, there is partial antigamy between the groups in alternate generations; but the intermingling is maintained because sons tend to maintain rights ceded to their fathers by their mothers' brothers in sites near their own patrilocal residences. Thus, they remain in constant contact with their mothers' natal groups and occasionally receive new grants of land on the basis of their maternal connections.

More intermarriage elaborates the web of cognatic relationships for descending generations and results in further intermingling of the garden sites of the two groups. People come to talk of the subterritories of the two groups as being one, and the border becomes indistinct. Eventually members of each group can trace cognatic connections to almost everyone in the other group. Cognatic connections replace marriage as the preferred means for acquiring rights in land and intermarriage ceases. The affinal cluster has become a cognatic cluster. Maring kinship terminology would facilitate this process. Kin terms are Iroquois on ego's generational level and bifurcate merging in the first ascending generation, but generational in the first descending and second ascending generations. Terminological distinctions between members of one's own exogamous group and groups with which it has intermarried are thus obliterated in two generations. Whereas the children of affines refer to each other by special cross-cousin terms, the children of cross-cousins refer to each other by the terms for brother and sister, although they may marry. This usage may be a formal concession to group dynamics: separate intermarrying groups can become a single corporate cognatic group, or cognatic cluster, without requiring a change in the kin terms that people of either group use to refer to the other.

It may further be that cognatic clusters become putative agnatic descent groups with the passage of time. Since a subterritorial clan cluster is functionally equivalent to a subterritorial clan such a transformation requires only a change in "charter." Once again their kinship terminology would facilitate such a change, since it masks the distinction between cognates and agnates after two generations. Ideological consistency would call for such a transformation, for in the Tsembaga view territoriality is ideally associated with agnation. It might be expected, a priori, that when knowledge of actual connections fade agnatic connections between the constituent (clan) units of the (cognatic) cluster will be assumed. This expectation is not, however, supported by the cognatic cluster found among the Tsembaga. A marriage connection between the KwiBigai and the Tsembaga clan is recent and therefore well remembered. In the case of the Tsembaga clan and the Tomegai, however, which have shared the subterritory for a much longer time, specific marriage connections are no longer remembered but the two clans retain their separate agnatic identities.

So far I have only spoken of land amalgamation leading to group fusion. This process is sometimes reversed, however, as may be illustrated from the tradition of the easternmost of the three subterritorial groups, the Merkai. At the time of my fieldwork this group consisted of a single exogamous clan but in the past, before they and the other four clans joined together as the Tsembaga local population, members of one of Merkai's three subclans began to intermarry with members of the other two. Informants say that when the clan was far more populous and occupied less land than it does at present one of the subclans moved its houses away and began to concentrate its gardening activities in the vicinity of its new settlement. Eventually its lands became separate, and intermarriage with the other two subclans began, although common ancestry was still acknowledged.

The process of fission seems later to have been reversed and it may be suggested that this reversal was correlated with a reduction in population pressure. After intraclan marriage began, the Merkai both won additional land and experienced a reduction in numbers as a result of disease, warfare, and emigration. At any rate, in 1963 the lands of the Merkai subclans were again intermingled. However, one intraclan marriage had taken place in the early 1950s, long after any population pressure that the Merkai had been experiencing had been relieved. Informants say that this marriage caused considerable consternation and that no further intraclan marriages would take place. This statement reflects, at least, demographic probability, for the Atigai subclan, which has in the past intermarried with the other two, has been reduced to seven persons.

Returning to the process of fission, what may have induced the removal of the Merkai Atigai residences is beyond the memory of any Tsembaga, and I will refrain from speculating about the mechanisms by which land segregation, if it actually occurred, may have been effected. All that is certain is that intermarriage between what are presently subclans of a single clan did take place, and that in former times a larger population occupied a smaller area.

While it is not possible to be certain, there is strong reason to believe that the Merkai were approaching or exceeding the carrying capacity of their land during at least part of the period when intermarriage was taking place among what are presently its subclans (see pp. 114ff.). Although data based upon my own observations or the observations of living informants are lacking, I believe that informants' statements relating, at least temporally, the transformation of antigamous subclans into intermarrying clans with population pressure should be taken to reflect a systemic possibility if not an historical fact. This belief is reinforced by Meggitt's observation of a similar process among the Enga people of the Western Highlands (1965:16).

In this connection it may be noted that although the practice of conveying land between members of different subclans of the same clan and between members of different clans or subterritorial groups may disperse population evenly over available land, it cannot alleviate a general land shortage. As critical density is approached or exceeded it may be that land-claiming units at all levels discourage the cultivation of their lands by members of coordinate units. An increasing emphasis on the primary principle of filiation, in this case agnation, as the basis for claims to land might then be expected, as Meggitt (1965:260ff) has proposed. Intraclan marriage accompanied by affinal land transfers might well be part of the associated strategy. Such marriages transform clan agnates into affines (affinal, rather than agnatic terms of reference and address are used), and affinal transfers of land may replace the recipients' residual rights as agnates in the subclan estate of the donors with the more limited rights of affines in specified garden sites, while leaving intact or ambiguous the donor's agnatic rights in the lands of the recipient. Since this post hoc interpretation did not occur to me until after I left the field, sufficient data to support or refute it are lacking.

Vayda and Cook have suggested concerning social organization in New Guinea that "it is probably more rewarding . . . to focus upon . . . processes than to attempt to devise elaborate social typologies applicable to all the structural variations encountered upon a single time plane" (1964:802). It may be suggested that the formal structure of the Tsembaga local population at any point in time is a more or less ephemeral product of continuing processes of population dispersal over available land, and that the discussion here bears upon the more general discussion of the relationship between population pressure and agnation in Oceania that has been carried on over the past decade (cf. Brookfield and Brown, 1963:170ff; Goodenough, 1955:80f; Meggitt, 1965:260ff). Tsembaga data would suggest that when population densities are low and considerable distances separate neighboring groups there is little permanent intermingling of garden land only because land acquired by virtue of membership in an agnatic group is convenient to a man's residence and other lands are not. With moderate to moderately high densities and reduced distances between neighboring groups, conditions that currently prevail among the Tsembaga, intermingling increases, for conveniently located lands may be obtained through affinal and then cognatic connections. As lands become intermingled, previously distinct social groups fuse, and in doing so they become antigamous. As critical density is approached the primary principle of filiation is emphasized as the basis for rights in land, garden intermingling is inhibited, and social groups are likely to fission or split and may begin to intermarry. This interpretation tends to support Meggitt's argument concerning the relationship between agnation and population pressure in the New Guinea Highlands, but it also reconciles some of the differences between him and Brookfield and Brown.

Clan histories indicate that warfare played an important part in the processes through which the five agnatic descent groups composing Tsembaga emerged as a single distinct unit from a regional population composed of many agnatic descent groups dispersed more or less continuously over the entire Maring area. People say that approximately fifty years before my fieldwork the gardens of the Kamuŋgagai, the westernmost Tsembaga clan, and the Kundagai to their west were intermingled, and litigation that has been initiated since the imposition of government-enforced peace supports their memory. But in a fight over a woman one Kundagai was killed and warfare immediately followed. Each of the embattled clans received, generally, the support of the clans behind it in this encounter. The Kamuŋgagai thus received assistance from the Tsembaga, Tomegai, and Kwibigai, with whom their garden lands were intermingled and with whom they had intermarried frequently, and from many Merkai, with whom they had married less frequently. They also received some support from the Dimbagai-Yimyagai immediately to the east of the Merkai, and the Tuguma local population, which occupied the territory beyond the Dimbagai-Yimyagai. Within days of the successful termination of this round of warfare trouble developed between the Merkai and the Dimbagai-Yimyagai.

In subsequent fighting the Merkai prevailed, with the support of most of the Kwibigai, Tsembaga, Tomegai, and many Kamuŋgagai, and with the further cooperation of the Tuguma who had a grievance of their own against the Dimbagai-Yimyagai. The Dimbagai-Yimyagai were driven from their territory, part of which the Merkai eventually annexed. There is reason to believe that the fight between the Merkai and Dimbagai-Yimyagai, at least, was a response to population pressure (see pp. 114ff.), but whatever its causes warfare seems to have defined the borders of Tsembaga territory, and the Tsembaga as a unit were distinguished from other units by their joint participation in the fighting that defined these borders. This de facto association of previously autonomous units then became a de jure structure through the synchronization of the *rumbim* rituals that follow the successful termination of hostilities. These rituals will be discussed in detail in later chapters.

POLITICAL STRUCTURE

Some of the events to be discussed later depend upon the coordination of the activities of many persons; the means by which such coordination is achieved requires brief description.

There are no hereditary or formally elected chiefs among the Tsembaga, nor are there any named, explicitly political offices. Neither is there to be found a pattern, such as that described by Oliver (1955), in which certain individuals, having achieved the status of "big men," command or coerce the activities of subordinates and vie with each other in feast giving. Among the Maring, to be sure, some men are recognized as *yu maiwai* ("big," or "important" men) and are especially influential in public affairs. They do not, however, compete in feast giving and do not command the obedience of others. The ability of such a man to effect compliance with his wishes depends upon his persuasiveness, and not upon his exclusive occupancy of a particular position in the social or political structure. Indeed, there is no limitation upon the number of big men that may be present in any subclan or clan: the Tsembaga are truly egalitarian in that there are as many big men as there are men whose capabilities permit them to be big men. Moreover, there is not on the part of men in general any abdication, either expressed or tacit, of decision making in favor of big men. Everyone has a voice in decision making, if he cares to raise it, and anyone may attempt to initiate action by himself proceeding to act and thereby instigating others to follow.

If the term *authority* is taken to refer to a locus in communications network from which flow messages instigating actions, then it may be said that among the Tsembaga authority is shifting. Big men may perhaps be defined statistically: they are those who more frequently than other men initiate the courses of action to which a group commits itself. A man is not involved frequently in decision making because he is a big man; he is a big man because he is frequently involved in decision making. It should be made clear, however, that a big man is seldom under an obligation to participate in the making of any decision. He may, and frequently does, leave decisions to others. His motive for refraining from participation in a decision may be lack of interest, inability to arrive at an opinion, or the desire to avoid a thorny issue. But his right to silence is unquestioned, and its invocation, if not too frequent, does not diminish his status.

Whether or not a man is a big man depends upon his personal attributes. Big men, it is said, have "talk" (*čep*). They have talk concerning fighting, women, rituals, and gardens. They have opinions about things concerning the group, which they can express articulately and which their auditors respect. They are, in short, intelligent men of forceful personality. They are usually men of considerable physical strength and vigor as well.

Big men tend to be wealthy, tend to be shamans, and tend to be in possession of knowledge of the rituals concerned with fighting. Wealth and esoteric knowledge and ability are primarily the fruits of the same abilities—intelligence, vigor, forcefulness—that make a man a frequent decision-maker. While the possession of wealth and esoteric knowledge tends to support the status of a big man, it does not per se confer decision-making prerogatives. It does not, moreover, provide a particularly powerful set of tools that a man may use to bend others to his will, as may be the case in societies in which the difference in amounts held by the wealthy and the nonwealthy is great, or where esoteric knowledge surrounds its possessor with great sanctity. The "spheres of influence" of even the most respected big men are very limited. The ability of any such man to effect compliance with his wishes diminishes with structural distance: it is greatest within his own subclan and among the residents of his own men's house, less among other subclans within his clan and among residents of other men's houses, and even less (although perhaps still considerable) among members of other clans within the local population. The renown of such men usually transcends the local population, but their direct influence outside their own local group is restricted to affines, cognates, and nonkin trading partners.

The processes of decision making are as amorphous as the structure within which they take place. Meetings are sometimes called to discuss an issue, but these are rare. At meetings that I attended, furthermore, there was little attempt to reach decisions in any formal way. There was no one to frame propositions in the form of anything like motions that could be put to a vote, for one thing, and besides, the idea of voting itself is unknown. It may be suggested that formal decision making is actually avoided at meetings, for the framing of issues in terms that would allow a decision to be made could lead to confrontations between those holding opposing views. Such confrontations would be difficult to resolve. Meetings are simply events at which there is much discussion about a particular subject going on at the same place and time among an assemblage that is larger than usual. These meetings are strange in appearance. Small knots of men—three or four or five in a group—stand or sit on the ground and talk among themselves. There may be many such groups within a restricted area. A few men move from group to group. Occasionally someone will address the entire assemblage in a loud voice. Some men drift away, others come by. Eventually everyone drifts away. No decision has been reached and no action initiated, but there has been much talk. Concerning most action there is no meeting, but there is discussion in the men's houses and on the paths and in the gardens. A meeting crystallizes sentiment more quickly, but its purpose is limited to forming a consensus and not making decisions or instigating action per se. Its purpose is "to make the talk one" (reach agreement) more quickly than is usually the case. When some man thinks the consensus has been achieved, or when he thinks that there has been enough discussion, without further discussion he initiates the course of action that the consensus suggests: for instance, he puts his eel traps at a traditionally designated place in one of the streams, thus initiating preparations for the final stage of the *kaiko* or pig festival; he begins to collect materials required for the construction of visitors' houses on the dance ground; or he personally visits another local group and, through a kinsman, extends an invitation to the *kaiko*.

It frequently happens that he who attempts to instigate group action has misjudged the consensus and he is not followed. A young man named Borgai, for instance, placed his ritual eel traps in the water in late August 1963. No one else followed suit immediately; indeed, some men didn't place their traps until October, and by the time eels were finally used in November, most of those caught early by Borgai had died. But to find oneself without an immediate following is not the lot of young men of little account only, and to instigate action successfully even a recognized big man must sometimes persevere. In late 1962 there was much talk of the need to build a house on the periphery of the dance ground to shelter visitors to the pig festival, and one morning Yemp, who is both the government-appointed *luluai* and recognized by the people to be a *yu maiwai*, called out that construction would start immediately. Absolutely no one joined him. Undeterred, he proceeded to the forest and began to cut poles and gather *pandanus* leaves for the structure. He continued to work alone for three days, complaining bitterly to those who passed by about the worthlessness of Tsembaga men, their sole interests being gardening and copulating. On the fourth day others, perhaps out of shame or perhaps because Yemp had stimulated their sense of public duty, finally joined him in the work.

CHAPTER 3

Relations with the Immediate Environment

In the last chapter it was suggested that the Tsembaga are a population in the ecological sense since they, separate from neighboring groups, constitute a unit in a set of material exchanges with the populations of other species with which they share their territory. In this chapter Tsembaga territory, its climate, and biota will be described, and the material relations of the people with their nonhuman coinhabitants will be discussed.

THE ENVIRONMENT

Some measurements of the climate and descriptions of the biota of Tsembaga territory presented here will not enter the later analysis as variables. Nevertheless, they warrant attention. For instance, temperature and rainfall are discussed here not because the quantities themselves will be treated as variables, but because they represent conditions that may possibly affect the presence or absence of the various biotic components of the Tsembaga ecosystem. The cultivation of particular plants, for instance, is limited by temperature, rainfall, insolation, and soil conditions (Brookfield, 1964:20ff.; Kroeber, 1939). Information concerning climate, soils, and vegetation may thus constitute an important empirical basis for comparison between areas.

Climate

Between December 1, 1962 and November 30, 1963, 153.89 inches of rainfall were recorded at an altitude of 4,750 feet on Tsembaga territory. No figures from any earlier period are available. While there was a drier and a wetter season, recognized terminologically by the Tsembaga, in no month did less than 6.76 inches of rain fall. Moreover, three of the months that, according to informants, were supposed to be part of the dry season, August, September, and October, were very wet.

Rain generally fell gently, although on several occasions over 100 points fell in less than a half hour. The highest daily recording was 4.01 inches, and there were forty-three days when over 1 inch of rain fell.

Most rainfall occurred at night, and on most days sunshine was prolonged, although cloud was seldom absent from some part of the sky. From August through November, however, much of the rainfall occurred during the daylight hours, and days of prolonged sunshine were few. Evaporation, consequently, was impeded and the ground remained wet. Fog, technically cloud, is common, particularly above 4,000 feet. Although it appeared most usually between 4 and 7 P.M., it would occasionally persist through the day.

During the year there was sunshine on more than 243 days and rainfall on 253 days. Monthly rainfall figures for Tsembaga, and a summary of five years of rainfall statistics for Tabibuga in the Jimi Valley, are included in Appendix 1.

Seasonal variation in temperature is slight. The diurnal variation ranges from 7° to 16°F, with daily maxima almost always falling in the mid- to high 70s and the daily minima in the low 60s.

No winds of a force sufficient to break small twigs off trees were noted during the year, although Clarke (personal communication) reports that he experienced somewhat stronger winds in 1964. Informants say that they have never known wind to reach a force sufficient to damage houses, groves, or gardens.

Land

Tsembaga territory measured orthographically includes 2,033 acres, or 3.2 square miles, of which 1,690 acres, or approximately 2.5 miles, are in the Simbai Valley, while the remaining 343 acres are in the Jimi Valley. Population density reckoned against total orthographic area is thus about sixty-four persons per square mile.

The terrain is rugged, rising from an altitude of 2,200 feet at the Simbai River to 7,200 feet at the Bismarck Ridge. Slope is about

20° up to 5,000 feet, then rises more steeply to the mountain top. The surface is further complicated by spurs projecting at approximate right angles to the line of the ridge, and by frequent watercourses.

Information concerning soils, which seem generally to be poor, is summarized in Appendix 2.

Vegetation

Hundreds or even thousands of species of plants are to be found within the limited area circumscribed by the borders of Tsembaga territory. Native nomenclature is elaborate. In the case of most nondomesticated plants, the native taxon usually corresponds to a species designation, while among the cultivated plants elaborate distinctions are made at a subspecies level.

This rich flora is distributed over Tsembaga territory in several different plant associations, among which the Tsembaga make clear terminological distinctions, which may form the basis for our discussion here. These associations include:

1. *geni:* forest that is said never to have been cut
2. *korndo:* grassy areas devoid, or almost devoid, of trees, and dominated by *Imperata cylindrica*
3. *riŋgop:* associations of domesticated plants, i.e. gardens in production
4. *riŋgopwai:* secondary growth, of which there are two subtypes
   (a) *kikia:* secondary associations dominated by herbaceous species
   (b) *dukmi:* secondary associations dominated by woody species

Only *geni* and *korndo* will be discussed here; *riŋgop* and *riŋgopwai* will be discussed later in the context of subsistence activities.

GENI

Of the 1,690 acres held by the Tsembaga within the Simbai Valley, 602 by orthographic measurement are in virgin forest, which lies unbroken above a line varying in altitude from 5,000 to 5,400 feet. An additional 28 acres of virgin forest have survived as remnants at lower altitudes, and in the Jimi Valley the entire Tser holding of 343 acres is either in primary forest or very advanced secondary growth. Of the total area of 2,033 acres included by orthographic measurement in Tsembaga territory, 973, or 48 percent, are in virgin forest or in secondary growth resembling virgin forest.

Two different virgin forest associations may be distinguished on structural grounds, although no such distinction is recognized in the Maring language, so far as I know. These are the high forest and the moss forest.

High forest survives only in remnants between 2,000′ and 5,000′, but it is unbroken from approximately 5,000′ to 6,000′. Throughout the high forest, three fairly distinct arboreal strata support an abundance of epiphytes, lianes, stranglers, and small climbers. While there are two lower strata, one of shrubs and young trees, the other composed of low herbaceous forms, the forest floor generally appears open, and it is only in scattered locations that visibility at the ground is less than 100′.

There are some floristic differences between the upper and lower portions of the high-forest range, with the transition occurring between 3,500′ and 4,500′. The lower zone is referred to by the Tsembaga as the "*wora*," and the entire area above it is called the "*kamuŋga*." Information concerning floristic composition will be found in Appendix 3; it is enough to say here that all structural components of both the *kamuŋga* and *wora* high forest are extremely mixed. In a strip 200′x17′ at an altitude of 5,000′, the "A," or highest, tree stratum was found to include nine named tree types, the "B" stratum four, and the "C" stratum ten. Details concerning the forest census taken on this strip will be found in Appendix 3.

The most striking aspect of the high forest is the size of the trees. The crowns of "A" stratum trees achieve an estimated average height of close to 125′, with occasional individuals, particularly *nonomba,* a species of *Eugenia,* reaching 150′ or more. Girths are frequently massive; 8′ circumferences 3′ above the ground, or above the buttresses, which are a common feature, are not unusual. Trees of even greater girth are present.

Three-storied forests composed of trees of such size do not seem to be common at altitudes of 5,000 to 6,000 feet in most parts of the world (Richards, 1964). The forest observed at such altitudes on Tsembaga territory resembles lowland formations more closely than it does montane or submontane associations. The presence of this formation at such an altitude may be regarded as an example of "Massenerhebung Effect," the upward displacement of the altitudinal ranges of plant associations in inland regions as compared to coastal areas, and in the interior of mountain ranges as compared to foothills. Richards (1964:347) states that protection from wind is at least partially responsible for the Massenerhebung Effect, but Clarke in a personal communication suggests that patterns of cloud cover may also be important.

In the moss forest above 6,000′ trees are smaller, both in girth and height. The forest ceiling in this range is usually well under 100′ from the ground, and stratification is indistinct. Although scattered individuals reach heights of 100′ or more, their crowns do not form an unbroken canopy. Herbaceous epiphytes and climbers are less abundant than at lower ranges, but epiphytic mosses are much more luxuriant, completely covering the trunks of most trees to a depth of more than an inch. Information concerning the composition of the moss forest will also be found in Appendix 3.

KORNDO

In several parts of Tsembaga territory apparently stable grassy associations occur. Only one of these, located between 2,800 and 3,600 feet, is of considerable size, covering 41 acres. *Korndo* (*Imperata cylindrica;* p. e.: *kunai*), is the overwhelming dominant, although a few ground creepers are present, as are scattered tree ferns, most of which are *Cyathea angiensis*, and a few small thickets of a very large bamboo called "*waia,*" which have been planted. The oldest Tsembaga informants maintain that neither they nor their ancestors ever made gardens in this area, and that it has always been, as it is now, under *kunai.*

I have already mentioned that there is reason to believe that the Maring occupation of the Simbai Valley is relatively recent, perhaps having begun within the last 200 years. It may be that this association, if it is anthropogenic (Robbins, 1963), is the result of the activities of an earlier population for whose presence there is archaeological evidence in the form of the stone mortars and pestles that are occasionally found in the ground.

The sort of exploitation that would turn the cover over limited areas into *kunai* while surrounding areas remained in high forest is hard to visualize. It is certainly the case, however, that the frequent fires to which such associations are subjected encourage this grass, the rhyzomes of which are undamaged by fire and discourage most other plants.

Fauna

The nondomesticated fauna inhabiting Tsembaga territory is abundant. Placental mammals include only feral pigs, bats, and rats, of which there are ten or more named varieties. At least thirty named varieties of marsupials are present, and there are at least fourteen named snake types. Lizards and frogs are also represented by many named types, and the avifauna is very rich. The Tsembaga say that at least eighty-four named types are present on their territory. These include cassowary, several birds of paradise, many species of parrot, and "bush turkeys." Most of the carnivores are to be found among the avifauna, which include many owls, several species of hawk, and at least one eagle. In the streams both eels and catfish are found, although the latter is rare. Arthropoda are quite varied, and their nomenclature is elaborate.

The fauna includes few forms that imperil human beings. Five of the fourteen snakes are venomous. Only one of them, however, the *rarawa* (unident), is sufficiently poisonous, according to informants, to kill a human being, and the only instance of snakebite death that anyone remembers occurred many years ago, when an old woman succumbed. Wild pigs and cassowaries are capable of injuring seriously or even killing human beings, and gorings by pigs are not infrequent. Informants say, however, that neither wild pigs nor wild cassowaries are ever guilty of unprovoked attacks, and injuries are inflicted by them only when they are hunted.

Of greater danger to the welfare of the Tsembaga than any large animal are anopheles mosquitoes. Most of the Tsembaga suffer what are probably mild malarial attacks from time to time. Data are insufficient concerning intestinal parasites. Stools of only two persons, in addition to the Tsembaga fieldworkers, were subjected to laboratory analysis, and no parasites were found.

Spirits

The Tsembaga regard spirits (*rawa*) to be significant components of their environment, and the reasons they offer for performing many rituals that will be discussed later in this study concern their relations with them. To make the rituals comprehensible a brief introduction to several of the major categories of spirits is necessary.

In the *wora,* the lower part of the territory, a class of spirits called the "*rawa mai*" is said to dwell. The term *mai* appears in a number of other contexts, which illuminate both the usage here and the role the Tsembaga impute to these spirits.

A taro corm out of which rhyzomes have started to grow is a *mai.* A woman who has had a child is an *ambra mai,* and adult females of animals are *mai.* But femaleness is not necessarily implied by the term, for old men are *yu mai.* A meaning that seems common to all of these contexts is something out of which something else has grown.

The spirits of the low ground are concerned with growth and fertility. It is they who look after the increase and growth of people and pigs and the productivity of gardens and groves. They are concerned, too, with that portion of the fauna which inhabits the area below the unbroken high forest, which covers the land above 5,000 feet. Feral pigs are theirs, and when one is shot thanks must be given to them. Of particular importance to them are eels, which are said to be their pigs. As the *rawa mai* are concerned with the lower portion of the territory, so are they concerned with the lower portion of the body—the belly, the genitals, and the legs. Fecundity and strength in the legs derive from them, but so do afflictions of the belly and the groin.

The category of *rawa mai* includes two subcategories of related spirits. There are, first, those called "*koipa maŋgiaŋ.*" When informants are questioned they say that the *koipa maŋgiaŋ* of each clan is distinct, and that he dwells in a wide place in one of the streams on the clan's territory. Other contexts, however, suggest a single *koipa maŋgiaŋ* for all Tsembaga or even all Maring. A notion of separate manifestations of a single supernatural entity perhaps reconciles such inconsistencies.

Koipa maŋgiaŋ were never human, but near *koipa maŋgiaŋ* in "A" stratum trees in virgin forest remnants live the spirits of those Tsembaga who have died of illness or accident. They are referred to as "*rawa tukump.*" *Tukump,* in other contexts, designates the mold that develops on such articles as orchid fiber belts and the bark rope bindings of stone axes. These spirits, the "spirits of rot," are conceived as intermediaries between the living and *koipa maŋgiaŋ.*

Spirits of the low ground are said to be *kinim,* which in many contexts means simply "cold." Here, as in some other contexts, it carries an implication of wetness as well. The juice of sugar cane is *kinim,* as is water, and women are said to be so because of their vaginal secretions. Coldness and wetness are said by the Tsembaga to be conditions that together cause softness and rot, and softness and rot, in contrast to hardness and desiccation, are thought to be the necessary conditions for growing things. This is recognized in certain important rituals that will be discussed later; for now it is sufficient to say that the Tsembaga conceive the spirits of the low ground to be implicated in the cycle of fertility, growth, and decay to which all animate things are subject.

But in the Tsembaga view while decay is necessary to life it also implies death, and the spirits of the low ground are dangerous. The spirits of rot, particularly when newly deceased, may spread a kind of supernatural corruption, called simply "*tukump,*" which causes illness and harm to those exposed to it. *Koipa maŋgiaŋ* himself is particularly fearful. He alone among the spirits of local origin actually kills. Other of the local spirits may bring illness, but it is only *koipa maŋgiaŋ* who brings death, even when it is another spirit whose displeasure is responsible for such punishment. In sum, the notion of fertility and growth on the one hand and death and dissolution through decay on the other are linked in the persons of the spirits of rot and *koipa maŋgiaŋ.* The spirits of the low ground seem to be more than spirits of fertility; they are, rather, the spirits of a cycle in which life both terminates in and arises out of death.

In the *kamuŋga,* the upper part of the territory, a class of spirits called the "*rawa mugi*" (red spirits) reside. These are the spirits of Tsembaga who have been killed in warfare, and Tsembaga say that they derive their name from the fact that their deaths were bloody. Much of the moss forest and high forest near the top of the mountain is considered to be their home, and it is said that they have forbidden the felling of trees, except for certain ritual purposes, in this sacred area. The category of mammals termed "*ma*" (marsupials and perhaps some giant rats), which live in the *kamuŋga geni,* the virgin forest above 5,000 feet, are, moreover, considered to be their pigs, and when one is obtained in a trap or through hunting thanks must be offered to the red spirits. This is also the case when cassowary or other large birds are taken in the high-altitude virgin forest.

Aside from hunting and trapping in the *kamuŋga geni,* however, the red spirits are hardly concerned with the subsistence activities of the living. They are concerned, rather, with the relations of the Tsembaga to other local groups, particularly in the context of warfare. It is from the red spirits that the rituals associated with warfare are said to have been received by ancestors, and it is to them that they are mainly addressed. It is the red spirits, furthermore, who enforce the taboos concerning relations with the enemy during periods when hostilities are quiescent.

In contrast to the spirits of the low ground, who are said to be *kinim,* or "cold," the red spirits are *romba-nda,* or "hot." But as *kinim* in reference to the spirits of the low ground denotes wetness and softness, *romba-nda* in reference to the red spirits denotes dryness and hardness. While the cold, wet, and soft imply fertility, the hot, dry, and hard imply strength.

While both the hot and the cold, strength and fertility, are recognized to be qualities necessary to their survival and well-being, the Tsembaga regard them as opposing principles, inimical to each other. Certain activities therefore must be segregated from each other in time and space, and certain objects or persons must be prohibited from contact with other objects or persons. Men who have knowledge of rituals for propitiating the red spirits, for instance, are forbidden to eat snakes, because the "coldness" of these reptiles will damage the "hotness" of their rituals.

In further contrast to the spirits of the low ground, the red spirits are concerned with the upper, nonsexual part of the body: the chest, the head, and the arms, and afflictions of these parts may result from the displeasure of these spirits.

The red spirits are not the only supernatural inhabitants of the *kamuŋga.* Residing at *komba ku,* a limestone cliff at the highest point on Tsembaga territory, is a female spirit, or set of spirits, called the "*kun kaze ambra.*" *Yur kun kaze* is the term for the technique through which shamanistic ecstasy is produced. It involves smoking locally grown tobacco over which spells have been said, and singing ritual songs. *Yur* is the Maring term for tobacco. The meaning of *kun kaze* is obscure and is probably not Maring. It is the *kun kaze ambra* (*ambra:* woman) whom a shaman consults when he wishes to learn the will of the deceased.

In many contexts it would seem that there is a single *kun kaze ambra,* or "smoke woman," not only for all the Tsembaga but for all Maring speakers. In other contexts, however, it appears that there are many: as in the case of *koipa maŋgiaŋ,* the problem can perhaps be resolved in terms of a concept of local or even personal manifestations of a single supernatural entity.

While the smoke woman, who is said never to have been human, is conceived as female, this conception does not seem to carry any implication of fertility. She is not thought to be antagonistic or dangerous to women, but sexual activities must be segregated temporally from activities having to do with her.

Although she resides in the upper portion of the territory, unlike the *rawa mugi* the smoke woman is not associated with any animals or plants living on the high ground. Neither is she responsible for any portion of the human anatomy, although she is associated with the nose: it is through the nostrils that she enters the shaman's body. She is concerned, rather, with relationships among the spirits themselves and between the spirits and the Tsembaga. She must be contacted when a change from activities implicating the spirits of the low ground to activities implicating the red spirits is contemplated, and again when a change back is projected. Through her comes the consensus of the spirit world, approving, disapproving, advising delay. The smoke woman, in short, is thought to be a link between the living and the dead.

Only one other class of spirits needs to be mentioned here. These are *rawa tukump ragai.* These are the deceased of other clans. Unlike the Tsembaga's own *rawa tukump,* the term *rawa tukump ragai* includes the deceased of other groups who have been killed in warfare as well as those who died of illness or by accident. Particularly dangerous are the spirits of slain enemies, who are thought to lurk about the territory waiting for opportunities to avenge themselves by bringing illness or death to members of their slayer's group.

SUBSISTENCE

Tsembaga subsistence activities will be examined with a view to deriving values for some of the variables suggested in Chapter 1. Descriptions of actual subsistence procedures will form the subject of a separate work, and therefore are, to a large extent, excluded here.

Cultivation

Cultivation is considered to be the complex of behavior by which members of an animal species both propagate and care for the members of other species, either plant or animal, which, in turn, provide them with useful material. Among the Tsembaga cultivation includes horticulture, silviculture, and animal husbandry.

HORTICULTURE

The Tsembaga rely upon the products of their gardens for much the greatest part of their nutritional requirements. Their horticulture is shifting. It is a set of interrelated activities directed toward (1) the establishment and maintenance, in an area previously dominated by other plants, of a temporary association of cultivated plants in which edible species are most prevalent, and (2) the succession of the temporary association of cultivated plants by an association similar to that which it replaced.

The great preponderance of the gardens are made in secondary forest between 3,000 and 5,000 feet. Of the 381 gardens made in 1961, 1962, and 1963 that were surveyed or censused, only 1 had been cut in virgin forest.[1]

Informants say that some Maring people plant garden sites twice before abandoning them, but the Tsembaga plant them only once. Cropping continues for periods varying from fourteen to twenty-four or more months, after which the garden is allowed to revert to secondary growth. The major gardens are cut, for the most part, in the latter part of what is said to be the wetter season (April, May, and early June). Most burning takes place in the drier season

1. There are indications that other local populations of Maring speakers nearer to large expanses of low-lying virgin forest tend to make a somewhat greater percentage of their gardens in virgin forest.

(between June and September), and planting follows immediately. There is no calendar; the scheduling of gardening activities depends mostly on weather, although some attention is paid to a few indicator plants.

Men and women cooperate in making gardens. Both cut underbrush. Men alone fell and pollard trees, make fences, and lay out logs on the ground as plot markers, planting guides, and soil retainers. Either men or women may burn the refuse on the plot for the first time, but women are usually responsible for the second firing, which disposes of materials not burned by the first. The heaviest burden in planting falls upon women. Men plant bananas, sugar cane, and *maŋap* (*Saccharum edule;* p.e.: *pitpit*); women plant the tubers and greens. Harvesting also falls largely to women, as does weeding, although men sometimes help their wives in these operations.

Gardening pairs overlap; that is, a woman may, and usually does, during the same year, make gardens with more than one man. In addition to her husband, she will frequently make gardens with her own or her husband's unmarried brothers or a widowed father. Conversely, a man is likely to make gardens not only with his wife but also with his unmarried sisters or a widowed mother. It is not unusual, therefore, for a single individual to participate in making several gardens in a single season. An important factor contributing to the pattern of personnel arrangements in gardens is, of course, the disparity between numbers of males and females in particular clans or subclans.

If a gardening pair has many pigs, two kinds of gardens are planted during the dry season. In the lower altitudes, that is, between 3,000′ and 4,000′–4,500′, a *daŋ-wan duk* (taro-yam garden) will be planted, while in the upper altitudes, between 4,200′ and 5,200′, a *bo-ñogai duk* (sugar-sweet potato garden) is made. The native names for the two garden types indicate the crops that are most important in these gardens. Some sweet potatoes and sugar are found in almost all taro-yam gardens and vice versa. The differences in the composition of these gardens are mainly statistical, although some low-altitude crops—certain yams and hibiscus especially—are absent from sweet potato gardens in the higher altitudes.

When the number of pigs is small, separate sweet potato gardens are not planted. Instead, mixed gardens (*duk mai yant*), which bear closer resemblance in their composition to taro-yam gardens than to sugar-sweet potato gardens, are planted toward the middle of the altitudinal range.

In the wetter season, between November and April, most people plant additional gardens containing, for the most part, greens that come into maturity during the dry season. These gardens are always small and sometimes are not planted.

Information concerning the crop inventory is summarized in Table 2. Column 1 presents native categories, and column 2 provides

Table 2. Tsembaga *Riŋgop* (Cultivated Species Planted in Swiddens)

Native category	Identification, description, or criterial attributes	Popular name	Number of named varieties	Origin or introduction
1. daŋ-wan	"presentation tubers"			
daŋ	taros			
daŋ	*Colocasia esculenta*	taro	27	Present from time of ancestors, but new varieties appear, old disappear.
koŋ	*Xanthosoma sagittifolium*	taro kong	1	From Karam people in 1957.
wan	Yams			
wan	*Dioscorea alata* and *nummularia*	yam	32	Present from time of ancestors, but varieties change.
man	*D. bulbifera*	yam, mami	4	All varieties said to be present from time of ancestors.
diŋga	*D. pentaphylla*	yam	1	Said to be present from time of ancestors.
ruka	*D. esculenta*	yam	1	Said to be present from time of ancestors.
2. ñogai	"pig tubers"			
ñogai	*Ipomoea batatas*	sweet potato	24	Present from time of ancestors, but varieties change.
baundi	*Manihot dulcis*	manioc	1	Introduced from the Jimi Valley around 1920.
3. bep	Leafy greens not planted in pandanus groves			
čeŋmba	*Rungia klossi*		5	All said to be present from time of ancestors.
rampmañe	*Commelina* sp.		1	All said to be present from time of ancestors.

Table 2 continued:

Native category	Identification, description, or criterial attributes	Popular name	Number of named varieties	Origin or introduction
gonebi	*Brassica juncea*		1	All said to be present from time of ancestors.
rumba	*Cucurbitaceae* sp.		1	All said to be present from time of ancestors.
kiñipo	*Oenanthe javanica*	parsley	1	All said to be present from time of ancestors.
niŋk gonebi	?*Nasturium officinale*	watercress	1	Introduced from Jimi Valley, 1957.
4. komba-čem	crops planted in *pandanus* groves			
komba	*Pandanus conoideus*	*marita*	34	Present from time of ancestors, but varieties said to change slowly
čem	*Hibiscus manihot*	hibiscus	17	Present from time of ancestors, but varieties change.
bep	*Cyathea* spp.	fern	3	All present from time of ancestors.
5. bar	*Psophocarpus tetragonulobus* and *Dolichos lablab*	peas, beans	4	(P. tet) present since time of ancestors, (D. lablab) introduced from the Jimi Valley 1935–40.
6. yobai	*Musa sapientum*	banana	28	Present from time of ancestors, but varieties change.
7. konapa	*Zea mays*	corn, maize	2	Introduced from Jimi Valley, 1945–50.
8. maŋap	*Saccharum edule*	pitpit	16	Present since time of ancestors, but varieties change.
9. kwiai	*Setaria palmaefolia*	New Guinea asparagus	7	Present since time of ancestors, but varieties change.
10. pika	cucumber	cucumber		
pika	*Cucumis sativus*	cucumber	1	Said to be present from time of ancestors.
mop	?*Cucumis sativus*	cucumber	1	Said to be present from time of ancestors.
11. ira	*Cucurbita pepo*	pumpkin	1	From Jimi Valley 1945–1950.

Table 2 continued:

Native category	Identification, description, or crithrial attributes	Popular name	Number of named varieties	Origin or introduction
12. yibona	*Cucurbitaceae ?lagaenaria* sp.	gourd	1	?
13. raŋgo	*Zingiber c.p. zerumbet*	ginger	1	Present from time of ancestors.
14. bo	*Saccharum officinarum*	sugar cane	27	All said to be present from time of ancestors.
15. aramp	succulents eaten raw with salt			
kumerik	*?Pollia* sp.		2	Present from time of ancestors.
kiñkiñmai	*Hemigraphis* sp.		2	Present from time of ancestors.
16. punt	ornamentals			
amame	*Coleus seutellaroides*	green and yellow coleus	1	Present from time of ancestors.
nimp	*Coleus seutellaroides*	purple coleus	1	Present from time of ancestors.
korambe	*Impatiens platypelia*		1	Present from time of ancestors.
17. kañpamp	*?Broussonetia*	paper mulberry	1	Present from time of ancestors.
18. pai pai	*Carica papaya*	paw paw	1	Introduced from Karam people around 1940.
19. rumbim	*Cordyline fruticosa*	tanket	10	All varieties said to be present from time of ancestors.
20. yur	*Nicotiana*	tobacco	1	Introduced from Karam people during (?) 1920s.
Totals	Species: 36+		264+	

translations or meanings for these categories; in some instances English terms or Latin binomials serve adequately. In the case of some of the more inclusive categories (indicated by Roman numerals), this has not been possible, and notes on some of the entries are required.

Daŋ-wan, which translates "taro-yam," are designated "presentation tubers." *Daŋ-wan* figure in ceremonial food presentations, in contrast to *ñogai,* which do not. *Ñogai* are fed to pigs as well as people and are thus designated "pig tubers" here. The plants included in category IV, *komba čem,* are found in swiddens and are thus included here. They are also found separately in *pandanus* groves (see the discussion of silviculture below).

Information concerning the place of origin and time of introduction of the various crops is derived from informants' statements.

The tabulation of harvesting began on January 25, 1963, and daily records were kept for more than thirty gardens from February 16 to December 14, 1963. A schedule of harvesting by crop is provided in Figure 1. Until week twenty-four, this schedule represents a synthesis of observations on several gardens in which planting dates could be fixed within one week. After week twenty-four, however, planting dates could not be fixed with equal accuracy, for the plantings upon which this schedule is based took place before I arrived in the field. They may therefore be inaccurate by as much as two weeks.

It should also be kept in mind that there is variation in speed of growth between gardens that differ in altitude and in other respects. This diagram represents an approximation of the harvesting schedule in gardens at altitudes around 4,000 feet. At higher altitudes maturation is slower.

Quantities harvested per acre are presented in Table 3 for taro-yam gardens and in Table 4 for sugar-sweet potato gardens. These estimations, which are critical preliminaries to an estimation of the amount of garden land required to support the Tsembaga, are based upon the daily harvest records compiled in the field. The weights of major crop categories are presented individually, but caloric values for each, drawn from the literature (see Appendix 8), have been assigned so that there will be a degree of commensurability both between the various crops and between the totals and total figures from other areas in which the crop inventory is different.

These figures, which represent an attempt at estimating the yields of two major types of gardens from first harvesting to abandonment, should be accepted with caution. Because the period of fieldwork was considerably shorter than the period of harvesting, in both cases the figures represent a compilation of harvesting records kept for

Figure 1. Harvesting Schedule of Tsembaga Garden Crops

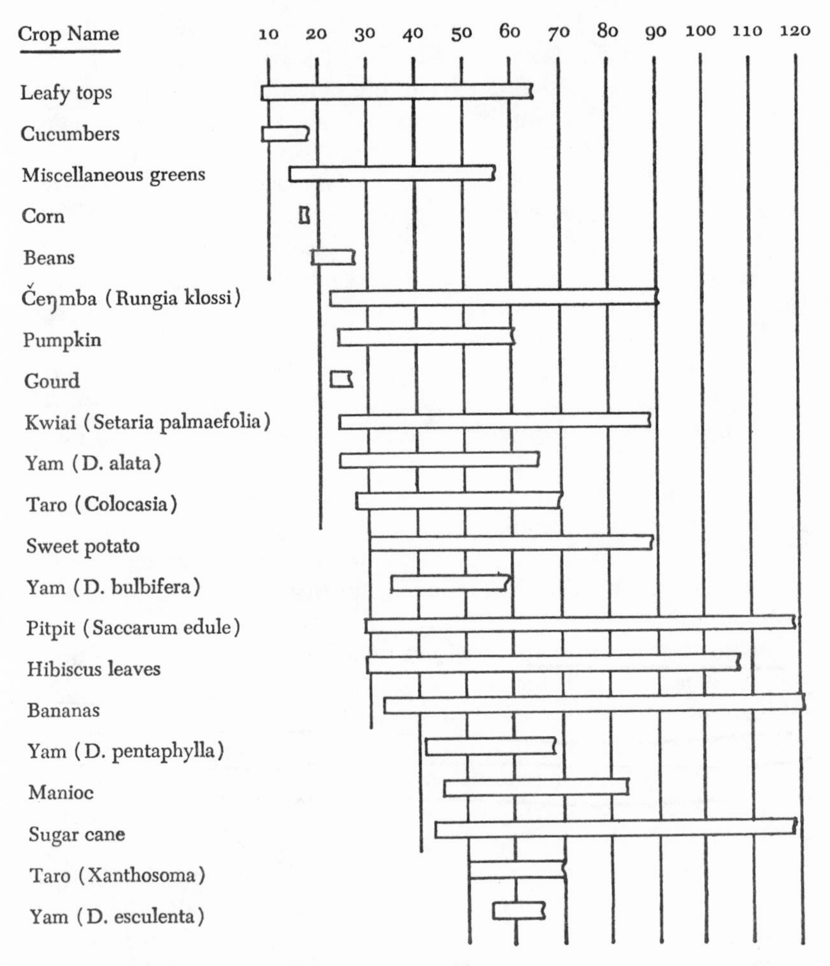

three gardens of different ages. For these and other reasons an error of 10 percent is not unlikely. The methods employed in these estimations are outlined in Appendix 4.

The per-acre yields of Tsembaga gardens are not high. Massal

Table 3. Yields per Acre—Tsembaga Taro-Yam Gardens

Crops	Yield in lbs. (0–23 wks.)[a] Torpai, 3.89[b]	Yield in lbs. (24–66 wks.) Kakopai, 3.3	Total lbs. (0–66 wks.) Torpai & Kakopai	Factor for nonweighed harvest, in %	Adj. total lbs. per acre (0–66 wks.)	Yield in lbs. (67–120 wks.) Tipema, 6.0	Factor for nonweighed harvest, in %	Adj. total lbs. per acre (67–120 wks.)	Total lbs. per acre (0–120 wks.)	Edible portion, in %	Calories per lb.	Total calories per acre
Colocasia	—	2,349.9	2,349.9	5	2,467.4	306.0	—	306.0	2,773.4	85	658	1,551,169
Xanthosoma	—	971.2	971.2	5	1,019.8	—	—	—	1,019.8	80	658	536,796
Dioscorea alata & *nummularia*	2.14	806.9	828.3	5	869.7	81.0	—	81.0	950.7	85	486	576,104 (*D. alata* & *nummularia*, *D. bulbifera*, *D. pentaphylla*, *D. esculenta* combined)
D. bulbifera[c]	—	188.9	188.9	5	198.3	18.0	—	18.0	216.3	85	486	
D. pentaphylla	—	159.3	159.3	5	167.3	4.0	—	4.0	171.3	85	486	
D. esculenta	—	53.6	53.6	5	56.3	—	—	—	56.3	85	486	
sweet potato	—	1,356.3	1,356.3	5	1,424.1	390.0	—	390.0	1,814.1	80	681	988,345
manioc	—	546.2	546.2	5	573.6	22.5	—	22.5	596.1	80	595	354,680
Rungia klossi	34.00	165.8	199.8	2	203.8	42.0	—	42.0	245.8	95	136	31,756
misc. greens	178.00	97.5	275.5	2	281.0	16.5	—	16.5	297.5	95	218	61,606
hibiscus leaves	—	636.9	636.9	2	649.6	120.0	—	120.0	769.6	95	136	99,430
beans	33.00	3.3	36.3	—	36.3	—	—	—	36.3	95	440	15,972
banana	—	534.9	534.9	20	641.9	822.0	20	986.4	1,628.3	70	427	486,685
corn	236.30	—	236.3	5	248.1	—	—	—	248.1	29	463	33,336
pitpit	—	459.8	459.8	2	468.9	26.6	—	26.6	495.5	40	104	20,612
Setaria palm.	33.00	468.9	501.9	2	511.9	610.5	—	610.5	1,122.4	17	65	12,402
cucumber	566.00	—	566.0	50	854.0	—	—	—	854.0	95	50	40,565
pumpkin	34.00	244.2	278.2	2	283.8	52.5	—	52.5	336.3	68	200	45,740
gourd	20.30	—	20.3	2	20.7	—	—	—	20.7	68	154	2,171
sugar	—	644.3	644.3	200	1,932.9	1,986.0	20	2,383.2	4,316.1	30	263	340,532
Totals									17,968.6			5,197,901

[a] Weeks after planting (entries in column 2, for example, indicate yields between 0 and 24 weeks after planting).

[b] The names (e.g., Torpai) are those of the sample gardens whose yields were weighed. The figures (e.g., 3.89) following the garden names are factors required to adjust the yields of the various sample gardens to yields per acre.

[c] *D.* = *Dioscorea.*

Table 4. Yields per Acre—Tsembaga Sugar-Sweet Potato Gardens

Crops	Yield in lbs. (0–23 wks.)[a] Torpai, 3.89[b]	Yield in lbs. (24–66 wks.) Timbikai, 4.2	Total in lbs. (0–66 wks.) Torpai & Timbikai	Factor for nonweighed harvest, in %	Adj. total lbs. per acre (0–66 wks.)	Yield in lbs. (67–120 wks.) Unai, 3.3	Factor for nonweighed harvest, in %	Adj. total lbs. per acre (67–120 wks.)	Total lbs. per acre (0–120 wks.)	Edible portion, in %	Calories per lb.	Total calories per acre
Colocasia	—	445.2	445.2	5	467.5	13.2	—	13.2	480.7	85	658	268,858
Xanthosoma	—	769.6	769.6	5	808.1	16.5	—	16.5	824.6	80	658	434,280
Dioscorea alata & nummularia	—	59.5	59.5	5	62.9	—	—	—	62.9	85	486	86,994 (bracketed for *Dioscorea alata & nummularia* through *D. esculenta*)
D. bulbifera[c]	—	38.8	38.8	5	40.7	1.6	—	1.6	42.3	85	486	
D. pentaphylla	—	70.3	70.3	5	73.8	—	—	—	73.8	85	486	
D. esculenta	—	—	—	—	—	—	—	—	—	—	—	
sweet potato	—	5,055.0	5,055.0	5	5,307.7	477.7	—	477.7	5,785.4	80	681	3,151,668
manioc	—	110.2	110.2	5	115.7	16.5	—	16.5	132.2	80	595	62,475
Rungia klossi	17.0	53.5	70.5	2	71.9	12.4	—	12.4	94.3	95	136	12,186
misc. greens	84.0	37.8	126.8	2	129.2	—	—	—	129.2	95	218	26,748
hibiscus leaves	—	40.9	40.9	2	41.7	26.4	—	26.4	68.1	95	130	8,799
beans	33.0	—	33.0	—	33.0	—	—	—	33.0	95	440	13,816
banana	—	45.1	45.1	20	54.1	293.7	20	352.3	406.4	70	427	121,482
corn	236.3	—	236.3	5	248.1	—	—	—	248.1	29	463	33,336
pitpit	—	405.3	405.3	2	413.4	316.0	—	316.0	729.4	40	104	30,264
Setaria palm.	16.5	266.2	282.7	2	288.4	29.7	—	29.7	318.1	17	65	3,510
cucumber	566.0	—	566.0	50	854.0	—	—	—	854.0	95	50	40,565
pumpkin	17.0	106.0	123.0	2	125.4	—	—	—	125.4	68	200	16,864
gourd	10.0	—	10.0	2	10.2	—	—	—	10.2	68	154	1,063
sugar	—	386.4	386.4	200	1,759.2	1,372.8	20	1,647.2	3,406.4	30	263	268,686
Totals									13,824.5			4,581,594

[a] Weeks after planting (entries in column 2, for example, indicate yields between 0 and 24 weeks after planting).

[b] The names (e.g., Torpai) are those of the sample gardens whose yields were weighed. The figures (e.g., 3.89) are factors required to adjust the yields of the sample gardens to per acre yields.

[c] *D.* = *Dioscorea*.

and Barrau (1956) present a number of estimates of yields of other types of agriculture found in the South Pacific. They state that mounded sweet potatoes produce three million to six million calories per acre within seven months, or a little more, after planting. If Okinawa variety sweet potatoes are planted, the yield might be as high as eight million calories (p. 25). Irrigated taro yields, they state, may reach eight tons per acre within a year of planting (p. 8). At values equivalent to those used for the Tsembaga swiddens, this represents close to nine million calories. Higher yields are also reported from the New Guinea Highlands. Brown and Brookfield (1959:26) report for Chimbu "long fallow cultivation" of sweet potatoes, and "comparatively high yields, certainly higher than the four tons per acre of sweet potatoes which Meggitt estimated in Mae Enga."

Tsembaga yields are not as high as these, but for purposes of comparison it should be emphasized that the great preponderance of the harvest of Tsembaga gardens is taken during the twelve-month period following the initial ripening of root crops. In the period between the twenty-fourth and the seventy-sixth weeks (approximately) after planting, Tsembaga taro-yam gardens yield crops containing 85.0 percent of their total of 5,197,900 calories. During the same period sugar-sweet potato gardens yield 90.7 percent of their total. It may also be noted here that calories are not, of course, all that gardens produce. Tsembaga gardens almost certainly yield larger amounts of plant protein, for instance, than do the gardens of the Chimbu and Enga.

An attempt has been made to estimate the amount of energy expended in producing an acre of both taro-yam and sugar-sweet potato gardens. This estimation is germane to calculations that will be made later in this study, particularly in reference to pig keeping and the timing of the ritual cycle. It also provides an additional dimension for evaluating Tsembaga subsistence procedures and comparing them to those in other areas. The results of these estimations are presented in Table 5, and a discussion of the methods employed in deriving the values will be found in Appendix 5.

Estimations of the lengths of the fallow periods for various gardens were difficult. The Tsembaga are not accustomed to reckoning in years, and in most cases it was necessary to arrive at dates of

Table 5. Energy Expenditure per Acre—Tsembaga Gardens

Operation (in sequence)	Expenditure value (kcals)	Derivation	Taro-yam garden	Sugar-sweet potato garden
Clearing underbrush	.65 cal/sq ft	Timing, H-K[a]	28,314	28,314
Clearing trees	.26 cal/sq ft	Timing, H-K	11,325	11,325
Fencing	46.17 cal/lin ft	Timing, H-K[b]	17,082	17,082
Preburn weeding	.1 cal/sq ft	Timing, H-K, modified	4,356	4,356
First burn	.008 cal/sq ft	Timing, H-K	336	336
Second burn	.11 cal/sq ft	Timing, H-K	4,792	4,792
Laying soil retainers	.168 cal/sq ft	Timing, H-K	7,238	7,238
Planting	.19 cal/sq ft	Timing, H-K	16,553	16,553
First weeding	.69 cal/sq ft	Timing, H-K	30,056	30,056
Second weeding	.69 cal/sq ft	Timing, H-K	30,056	30,056
Third weeding	.69 cal/sq ft	Timing, H-K	30,056	30,056
Tying sugar	1.80–3.60 cal/clump	Timing, H-K[c]	4,500	4,500
Misc. maintenance		(see note [d])	10,000	5,000
Acquiring supplies		Estimate for acquiring supplies, mainly lashing	8,500	5,000
Harvesting sweet potatoes	5.9 cal/lb	Timing, H-K	10,703	34,132
Harvesting taros	1.1 cal/lb	Timing, H-K	4,172	1,436
Harvesting manioc	3.0 cal/lb	Estimate, H-K	1,788	396
Harvesting yams	10.0 cal/lb	Estimate, H-K	13,930	1,770
Harvesting surface crops	1.0 cal/lb	Estimate, H-K	10,373	6,434
Walking to gardens	1.5 cal/min	H-K[e]	12,000	9,000
Carrying from gardens	6.56 cal/min	H-K[f]	59,404	39,360
Total			315,534	287,192

NOTES: Ratio of caloric return to input: taro-yam garden, 16.5:1; sugar-sweet potato garden, 15.9:1; and with carrying and walking reduced 80%, taro-yam garden, 20.1:1; sugar-sweet potato garden, 18.4:1.

[a] Hipsley and Kirk, 1965.

[b] Assumption of 370 ft of fence per acre.

[c] Assumption of 500 clumps per acre. First tie 1.8 cal/clump: second and third 3.6 cal/clump.

[d] Estimate for fence mending, yam house building, propping plants, etc.

[e] Allowance 20-minute walk down hill, 300 trips to sugar-sweet potato garden, 400 trips to taro-yam garden.

[f] 25-kg load carried up slope for 30 minutes, 300 trips from taro-yam, 200 from sugar-sweet potato garden.

previous cultivation by reference to events that themselves could be fixed in time only approximately.

The length of the fallow period varies between sites on the basis of a number of factors. These include local edaphic conditions and convenience to residences, but most important seems to be altitude. The full cultivation cycle measured from one planting to the next on sites between 3,500 and 4,200 feet is in some instances as short as ten years, although the average seems to fall around fifteen. At the higher altitudes fallow periods vary in length from an estimated twenty to forty-five years, with the average falling around twenty-five years. The figures of fifteen and twenty-five years, for lack of more precise estimations, will be used in later computations.

Estimation of the adequacy of fallow periods presents problems of criteria. The soil samples that survived the voyage to the laboratory cannot form the basis of a judgment; they indicate merely that fallow periods are necessary if soil depletion is to be avoided. The factors considered in estimating the adequacy of Tsembaga fallow periods are summarized here.

1. Several of the gardening practices that were observed seem to have conservation value. The Tsembaga practice selective weeding, for instance. From the time that weeding begins, five to eight weeks after planting, second-growth tree seedlings are allowed to remain, while herbaceous forms are uprooted. Not only does this avoid a definite grassy phase in the post-gardening succession, thereby minimizing the danger of a deflection toward a grassy disclimax, but it also provides during the cropping period deep tree roots that penetrate farther into the substrate than the roots of most crops. These roots are able to recover nutrients that otherwise might be lost through leaching. The development of trees over the garden also provides some protection for the exposed soil against tropical downpours and, furthermore, induces the gardeners to abandon the garden before the structure or content of the soil is severely damaged. People abandon gardens not because the crops are completely exhausted, but because the developing trees make the harvesting of the remaining root crops difficult.

The practice of pollarding large trees may also serve to protect the exposed soil during the cropping period. A fringe of leaves is left on the tops of many of the larger pollarded trees, and some of them survive their mutilation. Their roots may also recover some of the soil nutrients that are leached to levels deeper than may be recovered by the roots of the crops.

The practice of placing pigs in abandoned gardens for short periods (two to four weeks) may also contribute to the recovery of the sites. The pigs turn the ground, eliminating both herbaceous growth that has developed since the last weeding and some of the smaller seedlings, which if allowed to develop could provide root and light competition for already established saplings. The contribution of manure by pigs may also be considerable, but is probably offset by the organic materials they remove from the site.

2. Examination of the floristic composition of secondary growth at various stages of development also was relied upon as an indicator of the adequacy of the fallow periods. The absence or rarity of *Imperata cylindrica,* the grass that dominates disclimax areas on Tsembaga territory, on almost all garden sites suggests that deflection toward stable grassy associations is uncommon. On the other hand, the fact that in a census of twenty- to twenty-five-year-old secondary growth 54 percent of the tree species noted are also found in the virgin forest suggests that the secondary forest does not represent a serious deflection from the primary forest. Information concerning the floristic composition of secondary growth is summarized in Appendix 6.

3. The structure of the secondary growth in which gardens were being cut was also taken into consideration. In almost all instances the forest had formed an unbroken canopy over the site, and trees were of substantial size. There were in most plots many trees over 18 inches in circumference, and average heights were usually above 30 feet. The herbaceous species, while present, were for the most part limited to forms less than 2 feet in height.

4. Information from other areas, while not directly applicable, at least suggests the adequacy of Tsembaga fallow periods. Newton (1960:83), in reporting upon experimental work on shifting cultivation in Africa, states that the total quantity of nutrients immobilized within five years of garden abandonment is half that in eighteen years, and leaf development is also rapid. He states that the accumulation of litter on the forest floor reaches a maximum in eight to twelve years, and its soil-restoring decomposition and mineralization is rapid in tropical areas due to high temperature, humidity, and the prevalence of insects. There are, no doubt, differences between conditions in various areas, but if Tsembaga conditions are at all similar to those in which the experiments reported by Newton took place, Tsembaga fallow periods should be sufficient.

In consideration of these various factors, Tsembaga fallow periods were judged to be adequate; it is even possible that they are longer than necessary. In later computations, however, the values of fifteen and twenty-five years will be used.

SILVICULTURE

Tsembaga silviculture is a set of interrelated activities directed toward the establishment and maintenance of a permanent association of trees bearing edible materials in areas previously dominated by associations of other plants. Many kinds of trees are planted on Tsembaga territory; a list of the more important ones will be found in Appendix 7I. Attention will be confined here to the two forms that are planted in groves: *ambiam* (*Gnetum gnemon;* p.e.: *tulip*) and *komba* (*Pandanus conoideus;* p.e.: *marita*).

Ambiam, which is planted in groves below 4,000 feet, provides most importantly an edible green leaf. The fruit, which is known to the Tsembaga to be edible, is only infrequently consumed. This tree is less important in the diet of the Tsembaga than it is in those of many other Maring groups, which is at least in part a result of recent military defeat. After the Tsembaga were driven from their territory in 1953 their victorious enemies, the Kundagai, cut down most of their *ambiam* trees. The Tsembaga, however, are fortunate in having many other greens in their diet, many of which they prefer to *ambiam,* and replanting has not been heavy.

The Kundagai also cut down much of the Tsembaga *marita,* but heavy replanting has taken place since the Tsembaga began their return to their territory in 1957, and some men have as many as 700 trees in the ground. Unlike *ambiam,* for which there are many alternative greens, there is no substitute for *marita.* This tree bears a large waxy red or yellow fruit from which a sauce rich in vegetable fats is produced. Most of these trees were not yet bearing during the period of fieldwork, for *marita* planted on Tsembaga territory does not produce fruit until it is about five years of age.

No measurements were made of the amount of land under *marita* groves, of yields per unit area, or of the expenditure of energy involved in their planting or harvesting. The total amount of land under these groves is small, however, probably under fifteen acres, and is located for the most part below 3,000 feet. Some *marita* is also planted at higher altitudes, usually in gullies and ravines unfit for swiddening. Planting density is between 300 and 400 trees per acre.

The amount of work in planting and harvesting *marita* is slight, for the herbaceous cover need not be removed from the site, nor is fencing necessary. Harvesting involves simply knocking down the fruit with a long pole. Both planting and harvesting are men's work.

While no estimates were made of per-acre yields, the contribution of *marita* to the diet is significant, as reflected in consumption records (see Appendix 9).

ANIMAL HUSBANDRY

The Tsembaga keep four kinds of animals: pigs, dogs, chickens, and cassowaries (dogs were reintroduced during the period of fieldwork). All dogs previously owned by the Tsembaga had perished in an epidemic of a respiratory disease some years earlier. They were kept mainly for their usefulness in hunting and, in earlier times, for their teeth. Before shells were brought into the area forty to seventy years ago, the incisors of dogs and marsupials served as valuables in bride prices and other transactions. Some Marings, furthermore, eat dog flesh. Others are enjoined from doing so by taboos that not only apply to certain descent groups but also to people who are cognatically connected to those descent groups. Two of the five Tsembaga clans suffered such a prohibition against consuming dog flesh.

Some Maring groups keep large numbers of cassowaries; the Tsembaga do not. During most of the period of fieldwork, the Tsembaga had only one fully grown cassowary. Cassowaries do not breed in captivity. All tame birds are captured as chicks, and the Tsembaga usually trade cassowary chicks shortly after their capture to Jimi Valley people who, in turn, trade them across the Wahgi Divide, where they are important in bride prices. Among Maring groups cassowaries not only provided meat but also feathers, which were used in headdresses worn during fighting. Since pacification, cassowary feathers are not as valuable as they were. Chickens are few in number and hardly cared for. Rats kill many of the chicks, and hawks and other carnivorous birds carry off many of the grown chickens.

Pigs, in terms of their numbers, their contribution to the diet, and the effort required to keep them, are by far the most important of the domestic animals. Tsembaga ritual, moreover, like the ritual of many other people in Melanesia, is closely bound up with pigs. Most ritual occasions are marked by the slaughter of pigs and the consumption of pork. Furthermore, the timing of the ritual cycle and the occurrence of the year-long pig festival, called the *kaiko,* which terminates the five- to twenty-year cycle, depend upon the size, composition, and rate of growth of the pig herd. A more detailed discussion of the place of pigs among the Tsembaga is therefore necessary.

In June 1962, at the beginning of the *kaiko,* the Tsembaga herd numbered 169. At the termination of the festival in November 1963, the herd was reduced to 60 juvenile and 15 adult pigs. All of the surviving adults were scheduled for imminent killing, so that the herd was in effect reduced to 60 juveniles. This latter figure, moreover, is much higher than it would have been had I not been present. Wealth which I provided (by trading salt and beads for food, by paying young men in shillings for carrying supplies, etc.) was used for the purchase of a number of baby pigs that otherwise could not have been obtained. It is likely that in my absence the pig herd at the close of the *kaiko,* the point at which it is smallest, would have consisted of no more than 40 juveniles.

The average size of the animals also differed at the beginning and end of the *kaiko.* The average live weight of animals in June 1962 is estimated at 120 to 150 pounds. The average weight of animals surviving the *kaiko* is estimated at 60 to 70 pounds. The live weight of the herd was, if the surviving adults are discounted, about 5.6 times as great before the festival as after it. If a further adjustment is made to discount 20 of the juveniles present in November 1963, as a result of my presence, the difference becomes even greater. After such an adjustment the ratio of the live weight of the herd before to live weight after the festival is 8.4:1.

In addition to providing their owners with meat, pigs make at least two other contributions to the subsistence and physical well-being of the Tsembaga. It has already been mentioned that pigs confined in abandoned gardens not only utilize root crops that cannot be efficiently harvested by human beings, but also benefit the secondary forest that is developing on the site by uprooting much of its herbaceous component and thinning the arboreal component. Swiddens are sometimes planted twice by Maring groups in the Jimi Valley, and after the first crop is almost exhausted, pigs are penned in the gardens. Their rooting not only eliminates weeds and seedling trees, but it also softens the ground, making the task of planting for a second time easier. The pigs, then, are used by some Maring as cultivating machines.

Pigs make a further contribution to Tsembaga subsistence by eating garbage and human feces. Not only do they thus assist in keeping residential areas clean but they also convert wastes into materials that may be utilized by their masters.

Young pigs are treated as pets. As soon as it is weaned a baby pig begins to accompany its mistress to the gardens each day. At first it is carried. When it gets a little older it is led by a leash attached to its foreleg, but it quickly learns to follow its mistress in dog-like fashion and the leash is removed.

The young animal receives a great deal of loving attention—it is petted, talked to, and fed choice morsels. It shares the living quarters of the woman's house with the humans until it is between eight months and a year of age, when it is given a stall of its own. Even then it is not domiciled separately from its keepers, for stalls are inside the house, separated from the living quarters only by a rail fence through which the animal can thrust its snout for scratching or for morsels of food.

When it reaches four or five months of age the pig is considered old enough to look after itself and it no longer accompanies its mistress to the gardens. Instead, it is turned loose each morning to spend its day rooting in the secondary growth and forest and to return home in the evening, when it is given its daily ration of garbage and substandard tubers, mainly sweet potatoes.

This ration is substantial, but it probably does not comprise the largest portion of a mature animal's intake. The pig provides himself with most of his food in the course of his daily rounds. It is, however, an amount that is at least sufficient to induce most pigs to return home each evening and thus remain attached to the households of their mistresses. That is, it is an amount sufficient to keep most, but not all, animals domesticated in the face of considerable opportunity to become feral.

In this connection it may be suggested that as important to the pig as his ration of garbage and sweet potatoes may be, he may not return home for garbage and sweet potatoes alone. The work of Hendrix, Mitchell, and Van Vlack (1966) with horses indicates that certain kinds of human handling, notably petting and stroking, very early in a foal's life alter its relations with its mother and tie it closely to its human handler. For instance, foals that receive affectionate human handling do not follow closely behind their mothers as do foals that have not been so treated, but move considerable distances away. They also become unusually tractable and responsive saddle horses.

It is important to note that Hendrix et al. suggest that an important factor in the modification of the foal's development is the greater ability of the human handler (because he is equipped with hands) than the mare to provide tactile stimulation to the foal, and it may be that animals thus treated develop attachments to humans as strong or even stronger than attachments to their own species. Similarly, Lawrence K. Frank (1957) has observed that tactile communication is extremely important in the psychic development and socialization of human infants, and there are indications that this is true for the infants of other species as well (see Jay, 1963:119; Frank, 1957:201). It may be suggested that the petting and stroking to which Maring pigs are subjected as infants is an additional factor in keeping them domesticated throughout their lives. Such handling by humans communicates and produces positive affect, through which, along with his ration, the pig is bound to a social group dominated by humans. It is hardly facetious to say that the pig through its early socialization becomes a member of a Maring family.

Small numbers of pigs are easy to keep. To supply one or two animals with substandard sweet potatoes (under four ounces each) requires little extra work, for these tubers are taken from the ground in the course of harvesting the daily ration for humans. When the herd becomes large, however, the substandard tubers incidentally obtained in the course of harvesting for human needs become insufficient. It then becomes necessary to harvest especially for pigs—that is, to work for the pigs and perhaps to give them food fit for human consumption.

Records were kept of rations set aside for pigs at the four households of the Tomegai clan for a period of a little over three months (information concerning the sizes of these pigs is summarized in Table 6). The pig consumption figures were then extended to cover an additional five months for which harvesting and consumption figures were compiled for the Tomegai, and during which the pig and human populations were fairly constant.

Table 6. Roster of Pigs Owned by Members of the Tomegai Clan, 1962–63

Pig name	Size* March 1963	Date of acquisition	Size at acquisition	Month of death (1963)	Surviving size, Nov., 1963	Notes
Aŋgane	5			Nov.		
Gerki	5			Nov.		
Parau	4			Nov.		
Kombom	5			Nov.		
Tereŋ	5			Nov.		Bi-local pig; residing in Bank about 50% of time
Kikia	4			Nov.		
Koč Wai	3			Nov.		
Tambuŋ	2			Nov.		
Jokai	2			Nov.		
Bai	2				3	Scheduled for killing soon
Gi	2				3	Scheduled for killing soon
Grič	1				3	
Yuaneŋa	1				3	
Nameless		Aug.	1		2	
Prim		July	1	Sept.		Accidental death

* Pig sizes: (1) *wamba ñak* ("soft child"), under 40 pounds; (2) *wamba anč* ("hard child"), 40 to 80 pounds; (3) *baka* ("short"), 80 to 120 pounds; (4) *yundoi* ("large"), 120 to 160 pounds; (5) *yundoimai* ("very large"), 160 to 200 pounds.

The thirteen to fifteen pigs received considerably more sweet potato and manioc than did the sixteen humans. Of 9,944 pounds of sweet potatoes brought to the houses between March 11 and November 8, 1963, the pigs were given 5,554 pounds, or 53.7% of the total. Of 1,349 pounds of manioc brought home, the pigs received 1,106 pounds, or 82.0%. The total of all root crops brought home during this period in the four households was 18,574 pounds. The ration set aside for pigs was 6,674 pounds, or 35.9% of the total. An additional calculation suggests that an even larger proportion of root crops carried home eventually finds its way into the stomachs of the pigs. On the basis of observations (which accorded well with published figures on edible portion and wastage in root crops), 15% of the gross human ration of 11,900 pounds, or 1,785 pounds, may be assumed to have been wasted; that is, it was not eaten by humans. If half of this waste was peeling that was consumed by the pigs, the pig ration was 40.7% of the total root crops carried home. The humans received 54.4%, and 4.9% was wasted (burned in cooking, etc.).

It was not possible to weigh the size of the ration given to pigs of different sizes. Using as a basis figures of rations given to European pigs of different sizes under test conditions (FAO Agricultural Study #44:47), size 2 pigs receive about twice as much as size 1, and sizes 3, 4, and 5 receive three times as much. This calculation, which is gross and probably underestimates their rations, yields a figure of 2.6 pounds per day of tubers set aside for the adult and adolescent pigs. If garbage from tuberous portions of the human ration is added, the daily ration of adult pigs approaches three pounds, and it should not be forgotten that they receive other food as well, both other kinds of garbage and morsels from the rations of their masters and mistresses.

To provide such a ration for pigs is expensive. An indication of just how expensive may be derived from an examination of the differences in amount of land put in cultivation by the Tomegai in 1962, when they were supporting fourteen pigs averaging 100 to 130 pounds each, and in 1963, when they were looking forward to the reduction of the herd to three pigs averaging between 85 and 110 pounds. These figures are summarized in Table 7. The reduction

Table 7. Total Area of Tomegai Gardens in Square Feet

Garden type	1962	1963
Taro-yam gardens or mixed gardens	98,859	126,100
Sugar-sweet potato gardens	111,375	8,225
Totals	210,234	134,325

in land under cultivation immediately following the reduction of the pig herd by eleven animals was 75,909 square feet, or 36.1 percent. Since the difference is entirely in sweet potato gardens, and not in taro-yam gardens (out of which visitors to the pig festival are fed), it is reasonable to regard the reduction in garden area as a result of the reduction in the number of pigs being fed. It would seem that 1.65 acres were put into cultivation for the provisioning of eleven adult and adolescent animals. It may be mentioned here that the figure of .15 acres in cultivation per pig falls within the range, computed later in this chapter, for acreage under cultivation per person.

Another computation is of interest here. Figures derived earlier indicate that the energy required to produce 76,000 square feet of sweet potato garden is approximately 495,000 calories. The reduction in the herd was estimated at a live weight of 1,600 pounds at a maximum. This amount was estimated to produce, after dressing, about 800 pounds of uncooked flesh, or 50% of the live weight. At 1,318 calories per pound (FAO Nutritional Study #11:1954), this converts to 1,054,400 calories. The protein content of thin pork carcasses is taken at 10.9%, and fat 27% (FAO #11:1954), yielding for the ten pigs 87 pounds of protein and 216 pounds of fat.

It may not be necessary to point out that the ratio of energy derived from pork to energy expended in raising the pigs was *not* approximately 1,000,000:500,000, or 2:1. This would have been a poor enough figure from the point of view of energy efficiency. The actual ratio was even worse, for most of the pigs were two or more years of age when killed. The ratio could hardly have been better than 1:1 and may, indeed, have been even less favorable. That is, it is quite possible that more energy was expended to raise food for pigs than was returned in the form of pork. Pig raising, furthermore, involves energy costs other than those connected with raising food for them: lost pigs must be searched for, and damage to houses and fences by pigs must be repaired. The destruction of crops caused by invasions of pigs into producing gardens must also be regarded as an energy loss to the human population and, thus, one of the costs of pig keeping.

It is interesting to note that the energy ratio characteristic of the Tsembaga pig husbandry is less favorable than that of Siriono hunting, estimated by Harris (unpublished paper) to be 1.4:1, but probably closer to 2.5:1. Even this latter figure is not very favorable, and probably no human population anywhere could survive with such a ratio characterizing the energy-capturing activities upon which it relies most heavily. In addition to subsistence activities, other activities and basal metabolism must be supported by captured energy, and so must the activities and metabolisms of nonworking dependents: children, old people, and the sick and injured. It is little more than a guess, but I believe that very few techniques employed by human groups *primarily for the purpose of capturing energy* would show energy ratios much below 10:1. Less favorable ratios would barely cover the survival requirements of the population.

But it should not be assumed that energy-capturing activities are the only necessary subsistence activities. The survival and well-being of the human organism depend upon a supply of minerals, vitamins, and proteins as well as upon a supply of calories. In some cases the activity that yields these nutrients is the same as that which captures energy. Activities directed toward the production and harvesting of cereals, for instance, frequently fulfill at one and the same time both energy and protein requirements. But this need not be the case, and there is strong reason to doubt that the production and harvesting of root crops does so. It is not the purpose of this section to discuss the general adequacy of Tsembaga crops for the fulfillment of the protein as well as energy requirements of human beings. It is, however, important to emphasize here that while the details of energy transactions may illuminate some aspects of ecological and, perhaps, economic relationships, explanations that are restricted to the consideration of energy inputs and outputs will in some cases fail. We could only be mystified by a prolonged and laborious procedure, such as Tsembaga pig raising, which resulted in the return of somewhat less energy than was actually invested. We might, indeed, be tempted to construct theories that included among the advantages of the procedure only such empirically undemonstrable qualities as "mystic merit," or the prestige derived from conspicuous waste (Linton, 1955:98). On the other hand, if the frame is broadened so that not only energy input–energy output but also energy input–material output transactions are considered, activities such as Tsembaga pig raising are more clearly understood. The Tsembaga make an investment of energy and get a return of nutrients that are extremely important, but not primarily as a source of energy.

In light of the food and energy requirements of pig keeping and of the energy yield of pork, the applicability to the Tsembaga of a hypothesis recently advanced by Vayda, Leeds, and Smith (1961) concerning Melanesian pig keeping should be examined. These writers suggest that pig keeping may play an important part in the adjustment of Melanesian man to variations from time to time in the availability of foodstuffs, due to unfavorable weather.

> A way in which Melanesian populations are able to adjust to such unpredictable variation is through the practice of trying to plant more crops every year than can be or need to be consumed by the planters in a year not appreciably disturbed by adverse weather. Planting what will be more than enough should the weather be good is a means of ensuring that there will be enough . . . should the weather be bad. . . . If, then, the size of the human population is limited by the times of minimal yield, the vegetal surpluses of normal years become available for the feeding of livestock. . . . The practice of feeding vegetal surpluses to the pigs in the years of normal and maximal crop yields is described as "banking" in Oliver's account (1955:470) of a Solomon Island culture, and the term seems appropriate, for the pigs are indeed food reserves on the hoof. . . . In years of minimal yield when the garden produce has to be used entirely for the support of the local human population, attempts may be made to trade away or ceremonially to give away the pigs to distant and better-supplied communities or else to have the pigs agisted at such places. If these attempts are unsuccessful, the pigs must subsist on what they can forage for themselves. . . . Moreover, in the years of minimal yield, not only can the mature pigs in most cases take care of their own subsistence but they can also be a vital source of food for human beings. [1961:70, 71]

The mechanism outlined here does not seem applicable to the Tsembaga case. Weather bad enough to cause crop failures seems never to occur, according to informants. Indeed, they say that crop failures don't seem to occur for any reason. Anything less than a residence of many years' standing does not permit definitive statements concerning the occurrence or nonoccurrence of crop failures, but there is little reason in this instance to doubt informants' statements. They indicate that the most serious hindrance to cultivation suffered by the Tsembaga and their neighbors is excessive rain during the "drier season' (June–September), when most gardens are burned. Heavy rain during this period makes the task of burning more difficult but does not reduce the acreage placed in cultivation and has no effect upon yields. Field observations tend to support the informants' statements. July, August, and September 1963 were marked by unseasonably heavy rains, but I could detect nothing to suggest that the unfavorable weather resulted in less acreage in cultivation than would otherwise have been the case. No one, for instance, failed to burn gardens that had already been cleared of trees and underbrush. People simply worked harder at burning their gardens than would have been necessary if the weather had been drier. Informants say that comparably heavy rains had fallen during July, August, and September 1962. Very wet "drier seasons" apparently are not uncommon in the Simbai Valley.

Although crop failure caused by weather conditions may not have occurred, it is possible that food shortages could have developed out of the disruption brought by warfare. Informants denied this, however, and generally agreed that even during periods of active hostility garden work could be accomplished during frequent cease-fires.

The mechanism suggested by Vayda, Leeds, and Smith is inapplicable to the Tsembaga case for another reason as well. It posits an adjustment in which the amount of land put into cultivation by a population of constant size will remain more or less constant from year to year. In "normal years" this amount of land will yield food in excess of human requirements, and the excess will be fed to pigs, which may be regarded as "repositories for surplus vegetal produce" (1961:71). It has already been noted, however, that among the Tsembaga the amount of land put into production is not constant from year to year. The amount of land in production in any year depends upon the sizes of both the human and pig populations. As the pig population increases, so does the amount of land under cultivation. Pigs are not only fed from the surplus of gardens planted for humans: acreage may be put into production for the pigs.

The Tsembaga case does not provide a general refutation of the Vayda, Leeds, and Smith hypothesis, however. Their larger point, that Melanesian pig keeping is not wasteful, and that on the contrary "pigs are vitally important to the management of subsistence by

Melanesian populations" (1961:69) is supported. Although their more particular formulation does not fit the Tsembaga case, it deserves to be tested by quantified observations of pig husbandry, cultivation practices, and dietary intakes among Melanesian people living under a variety of weather conditions. It need hardly be said that in different ecological settings practices such as pig keeping should be expected to play different roles in the adjustments of their practitioners to their environments. It would not be surprising to discover that the mechanism suggested by Vayda, Leeds, and Smith corresponds closely to the situations in some areas subject to climatic catastrophe.

But in light of the discussion above concerning the capture of protein and energy, some comment must be made on one further point in the Vayda, Leeds, and Smith formulation: the characterization of pigs as repositories of vegetal surplus. The mere fact that pigs are fed domesticated plant foods not consumed by humans makes them, in a sense, repositories of vegetal surplus. The characterization suggests, however, whether or not it is the authors' intention, that the ration fed to the pigs is material that is in excess of the immediate physiological requirements of their keepers, and that it is merely stored in the animals for future use. The implication is that the material recovered from the animals is of the same sort as, or equivalent to, that which was fed to them at earlier times. In fact, the material recovered from pigs is quantitatively and qualitatively different from that which was fed to them, and it is in these differences that the main significance of the place of pigs in the diet of their keepers lies.

The ration fed to pigs consists mainly of carbohydrates in the form of sweet potatoes and tubers. These are most important sources of energy. However, animals are poor storehouses of energy since most of what they consume is expended in metabolism and activity, and it is unlikely that the amount of calories that can be reclaimed from the flesh of pigs raised by Melanesian practices would exceed one-fifth of the amount that they had received in rations.

It may also be recalled that pig husbandry is expensive in terms of the caloric expenditure demanded of its practitioners, and that the maintenance of pigs may also demand, as it does among the Tsembaga, the cultivation of acreage in excess of that required for the support of the human population. Where swiddening is practiced this could necessitate shortened fallow periods, which might produce changes in either soil structure or composition. These changes might well affect crop yields adversely. To put this in terms of the banking analogy, the "service charges" may be greater than the "savings account."

In view of the shortcomings of pigs as energy storehouses, and in view of the high cost of pig maintenance in both energy expenditure and land, it might well be asked if pig husbandry could be regarded as adaptive if the storage of energy for use in periods of crisis were the main advantage gained by those who practice it. The indications are that in many situations people would serve themselves better by giving up pig husbandry, letting surplus tubers rot in good years, and in years of shortage themselves consuming substandard tubers, which are edible even if undesirable. It may be suggested that only in environments in which even substandard tubers might become unavailable would the emergency energy supply provided by pork justify the high costs of maintaining pigs, and even in such environments pigs alone are of questionable value as a famine food. It is not likely that a Melanesian population would have sufficient pigs on hand to use pork as a substitute for carbohydrates until a new crop of tubers came in. The Tsembaga pig herd, for instance, at the outset of the *kaiko,* probably represented about 13,000 pounds of pork. A minimum of 1,000 to 1,400 calories a day per average person would be required to permit people to carry out even minimal activities. If such a ration had to be provided to a population the size of the Tsembaga by pork alone, the Tsembaga herd would be exhausted in two to three months. It is, of course, unlikely that no other foods would be obtainable, but it is also unlikely that such a quantity of pork would be available during a period of hardship to groups similar in size to the Tsembaga. This quantity reflects maximal herd size; usually the herd is smaller.

Although the advantages of pigs as storehouses of energy are dubious, the animals do render an obvious nutritional service to their keepers. The qualitative differences between the composition of pig fodder, mainly sweet potatoes and manioc in the Tsembaga case, and pork make this apparent. The edible portion of a lean pig yields 10.9% protein and 27% fat by weight (FAO Nutritional Study

#11:1954). The edible portion of sweet potatoes yields 0.9–1.7% protein and 0.3% fat by weight (Massal and Barrau, 1956). Manioc yields 0.7–1.2% protein and 0.3% fat (Massal and Barrau). While the amino acid ratios of sweet potatoes are quite good, those of manioc are extremely poor, and not all of the limited amount of protein that manioc contains can be metabolized by humans except in the presence of other foods rich in the specific proteins that manioc lacks. The disparity, thus, between the protein contents of manioc and sweet potato on the one hand, and pork on the other, is even wider than indicated by the figures. The difference between the fat contents of the root crops and the pork is in the ratio of 89:1. It seems, therefore, that rather than being viewed as a means for storing vegetables, pig husbandry might better be regarded as a means for converting carbohydrates into high-quality protein and fat.

In a later section the role of pork consumption at times of crisis will be discussed. It is sufficient to note here that a ready supply of high-quality protein is of considerable importance to the well-being of the Tsembaga and no doubt other Melanesian populations as well. Melanesian pigs, as Vayda, Leeds, and Smith point out, cannot be regarded as luxuries. They are a very expensive necessity. The lengths to which the Tsembaga go to maintain their animals is striking, but perhaps not extraordinary for horticultural people living on tuberous staples.

It is not only in respect to the amount of work entailed in their care that large numbers of pigs may be, as Vayda, Leeds, and Smith (1961:71) have pointed out, too much of a good thing. This is also true of the sanitary and cultivating services rendered by pigs. A small number of pigs is sufficient to keep residential areas clean and is also sufficient to suppress herbaceous growth and superfluous seedlings in abandoned gardens. A larger herd, on the other hand, may be troublesome—the greater the number of pigs, the greater the possibility of their invading producing gardens, with concomitant damage not only to crops and young secondary growth, but also to the relations between the owners of guilty pigs and the owners of damaged gardens. Furthermore, while exposure to a moderate number of pigs for a limited time directly after the abandonment of gardens may benefit the development of the arboreal component of the fallow by eliminating seedlings that might compete with established saplings, the prolonged exposure of a site to a large number of pigs is detrimental; many seedlings that, being at some distance from already established specimens, could grow to fill gaps in the canopy are constantly eliminated. The establishment of quick-growing, sun-loving herbaceous species, including *kunai,* is thus encouraged. The latter grass, when it becomes dominant, tends to form a fairly stable disclimax, its rhyzomatous subsurface structure making it difficult for other species, arboreal or herbaceous, to gain a foothold.

The effects of concentrating a large number of pigs in a small area could be observed around the Tsembaga settlement. It has already been mentioned that in the past the Tsembaga settlement pattern has been "pulsating," with residential units being of clan or subclan size most of the time. Not only people but also, and perhaps more importantly, their pigs have usually been scattered over the countryside. However, out of fear of their enemies the Tsembaga, upon returning to their territory in 1957 after a three-year exile, built all of their houses in a single nucleated settlement, which they continued to occupy in 1963. For six years 200 persons, rather than the usual 20 to 40, thus resided within an area 200 to 400 feet wide and about 2,000 feet long. Each day about half of their pigs passed through the grounds named "*Pra*" and "*Gerki*" just west of the settlement, rooting on their way. These areas were cultivated in 1957, and stumps bear out informants' statements that the sites were in secondary forest previous to that time. But six years later trees were only scattered. No canopy had formed, for the pigs had constantly rooted out the seedlings necessary to its development. *Kunai* seemed to be the most important component of the ground cover.

Usually such concentrations of pigs were not allowed to develop. In earlier times the extent to which settlements were scattered was, it seems, directly correlated with the size of the pig herd. During and directly after the pig festival the degree of nucleation was high. As the pig herd grew, however, people would move away, establishing themselves in hamlets. Even these sometimes disintegrated into scattered homesteads as the number of pigs grew even larger.

Factors other than pigs, of course, provided stimulus for this scattering. The Tsembaga and other Maring are likely to move their residences because of illness, which is thought frequently to be the result of the contamination of the ground either by supernatural *tukump,* corruption disseminated by a spirit, or by *kum,* a similar sort of areal infection sent by sorcery. But disease among pigs is frequently rationalized in precisely the same way and is probably as frequent a reason for moving as is disease among people.

Arguments concerning pigs also result in removals to isolated locations. These frequently have to do with garden invasion. Residential isolation serves to increase the distance between one's pigs and the gardens of others and the pigs of other people and gardens of one's own, thus diminishing the possibility of damage to both gardens and social relations. This is explicitly stated by some Tsembaga in explaining the location of their residences. Residential removal, usually resulting in the scattering of house sites, is also stimulated by the desire to decrease house-to-garden distances, as has already been mentioned, and the meaning of this in terms of energy expenditure has been presented. The need to reduce house-to-garden distances obviously becomes increasingly compelling with an increased pig herd, for it is necessary to transport food for pigs as well as people from gardens to houses.

While the process of residential "denucleation" mitigates the problems resulting from large concentrations of pigs in limited areas, it also undoubtedly results in less social interaction between members of different small residential groups. Further, the scattered residential pattern opens the group to greater danger from enemy action during periods of open hostility. In short, the Tsembaga pay for their pigs not only in land, food, and energy, but also in loss of opportunities for social contact and in increased vulnerability.

Certain breeding practices inhibit the rapid growth of the pig population. All male pigs are castrated at approximately three months of age. Tsembaga say that this practice produces a larger and more docile animal. It also obviously results in a reliance upon feral males for the impregnation of domestic sows. Matings must be infrequent since most of the sows are domiciled above 4,000 feet, while the feral males are inclined to stay below 3,000 feet. During the period of fieldwork, out of a potential one hundred pregnancies (sows can bear two litters a year) only fourteen litters resulted.[2]

Birth generally takes place in the forest and infants are sometimes

2. I believe that this figure is particularly low and should not be taken to represent the birthrate of Tsembaga pigs over longer periods of time. I suggest that the

lost. Perhaps some of those lost survive in a feral state, if so, they may serve later to impregnate domestic sows or may be recovered through hunting, but most of them probably perish.

Mortality among those infant pigs who do find their way home is also high. From the fourteen litters only thirty-two of the offspring were alive at the end of the fieldwork. If the seven littermates born in early December 1963 and still surviving at the end of the fieldwork two weeks later are excluded from the sample, an average of only two animals per litter survived infancy.

Hunting and Gathering

A wide variety of nondomesticated resources, obtained from all of the biotic associations found on the territory, are utilized by the Tsembaga. Information concerning the use and site of occurrence of several hundred of these items is summarized in Appendix 8, but the lists are far from complete.

Reliance upon any single item is not particularly heavy. While no quantitative data on the use of such materials were compiled, it is clear, however, that reliance upon the aggregate of nondomesticated resources is considerable. Of particular importance are firewood, game, and building materials. The land under primary and secondary forest should not be considered to be merely a reserve that could be put into production should the need arise, or land out of production during the fallow period. Such lands constitute important resource areas in their own right. Increase in the amount of land under cultivation not only might damage the environment by shortening fallows or encouraging erosion in the steep higher zone covered by primary forest; it would also reduce the amount of nondomesticated materials available to the Tsembaga.

THE DIET

Estimation of the food intakes of the Tsembaga are necessary to calculations that will be made later. One of the goals of this study

---

birthrate among the pigs is inversely correlated with the degree of residential nucleation. Feral pigs tend to avoid areas that are densely populated by humans, and domestic animals do not seem for the most part to wander very far afield. Moreover, in 1963 the movements of domestic pigs were restricted by the situation of the Tsembaga settlement, which was separated from much of the territory by streams the pigs found difficult to cross. With a more dispersed settlement pattern there would be considerably more opportunity for feral males and domestic sows to meet.

is to show how ritual operates to keep the trophic demands of the Tsembaga and their pigs within the carrying capacity of their territory. To do this, it is necessary to calculate the trophic demands of the population. In most estimations of carrying capacity, this quantification is expressed in terms of acreage per capita per annum put into production. Such a figure does not indicate, however, whether a per capita daily intake of 1,500 or 2,500 or 3,000 calories is allowed for. Estimations of actual intake specify the level at which the individuals who constitute the population are being supported, which quantification in terms of area under cultivation alone does not. Actual intakes, further, may be assessed against estimated requirements and a judgment can be made as to whether the population is being maintained in a state of well-being or at bare subsistence level.

Estimations of intakes, in addition to supplementing carrying capacity calculations, will form one of the bases for another estimation that will be of considerable importance in elucidating the homeostatic function of ritual: the amount of energy available for pig husbandry. The ways in which ritual and taboo tend to optimize the utility of the limited amounts of animal protein available will also be described.

Composition of the Diet

The composition of the Tsembaga diet by percentage of constituent foods by weight is presented in Table 8, along with comparative information from other New Guinea groups.

The data from Busama, Kaiapit, Patep, and Kavataria were collected during the New Guinea Food Survey of 1947. The broad nature of this survey did not permit long observation of consumption in any location, and it is probable that some items included in the diets of these communities were not recorded. It may be, therefore, that the impression of greater variety in the Tsembaga diet is to some extent misleading. It is worth noting, however, that the Tsembaga seem to rely less upon any single starchy staple than do other groups. Furthermore, a much larger proportion of the diet apparently consists of foods other than starchy staples. Slightly more than one-third of the Tsembaga diet by weight, 34.9%, is composed of leaves, stems, and fruits other than banana. Sugar cane (which will be included in later calculations of Tsembaga intake)

Table 8. Composition of Diets of Tsembaga and Other New Guinea Communities, by Percentage Weight

	Busama	Kaiapit	Patep	Kavataria	Chimbu	Tsembaga
Taro	65.0	7.3	45.9	8.6	—	25.8
Yam	1.4	9.5	.2	55.2	5.0	9.3
Manioc	—	2.0	.2	1.4	—	1.2
Sweet potato	—	25.7	37.6	14.0	77.0	21.0
Sago	6.8	.28	—	—	—	—
Banana	1.1	31.6	.4	5.5	—	7.8
Fruits and stems	6.1	—	2.6	2.3	13.0	17.3
Leaves	14.0	9.2	8.0	—	2.5	9.9
Coconut	2.2	9.5	—	2.8	—	—
Marita pandanus	—	—	?	—	?	4.2
Grain	.36	—	3.0	—	1.5	—
Misc. veg.	—	3.4	1.8	1.4	—	2.5
Animal	2.9	1.7	.2	9.7	1.0	1.0

SOURCES: Information for Busama, Kaiapit, Patep, and Kavataria from Hipsley & Clements, 1947. Information for Chimbu from Venkatachalam, 1962.

NOTES: Busama is a coastal village on the Huon Gulf. Kaiapit is 6 miles east of the Markham River, about 60 miles northwest of Lae, altitude about 1,000 feet. Patep is about 25 miles from the coast at the mouth of the Buang River, altitude 3,550 feet. Kavataria is a coastal village on Kirawina in the Trobriand Islands. Chimbu is at an altitude of 5,000–7,000 feet in a central mountain range.

has been excluded from this comparison since it is not included in the reports from some of the other communities in which it may be present. If it were included here, the proportion of items other than starchy staples in the Tsembaga diet would approach 50%. In the five other communities with which the Tsembaga are compared, the percentage of nonstarchy items in the diet varies from 15.3% to 25.7%. In further contrast to the others, the Tsembaga diet includes an appreciable amount of *marita* pandanus, the edible portion of which contains 14.0% fat, according to Hipsley and Kirk (1965:38). The nutritional consequences of the greater variety found in the Tsembaga diet will be discussed later.

Tsembaga Intakes

Assessment of the Tsembaga diet is based upon both the quantity of food ingested and its nutritive value. Vegetable and animal foods are treated separately.

PLANT FOODS

Fruits and vegetables comprise approximately 99 percent by weight of the usual daily intake of the Tsembaga. All vegetables brought home to the four hearths of the Tomegai clan were weighed daily by named variety, from February 14 to December 14, 1963. The figures from March 11 to November 8, a period during which both the pig and human populations were relatively stable, formed the basis of the estimations of intake. Details of the methods employed will be found in Appendix 9.

Estimated daily intakes of calories, protein, and calcium for individuals in various age and sex categories are summarized in Table 9, where they are compared to recommendations of both the New Guinea Food Survey (Langley, 1947:134; Venkatachalam, 1962:10) and FAO/WHO (World Food Problems #5, 1964:6).

In comparison to these recommendations, caloric intake for all categories of the population is adequate, but the calcium portions of the younger groups are not. These possible deficiencies are moderate in the case of children three to five years old and five to ten, while the lack possibly suffered by infants up to three years old is made up in whole or part by mother's milk, since children are nursed usually for two years or more.

In a discussion of protein intake, the quality of the protein must be considered. Foods vary in "biological value," that is, in the extent to which they may be utilized for maintenance and growth (Burton et al., 1959:46). This is a function of the proportions in which the various amino acids of which the protein is composed are present: some may not be utilized except in the presence of others, while excesses of some may inhibit the utilization of others (FAO #16, 1957:23).

It is generally the case that proteins of animal origin are "complete," that is, they contain appreciable amounts of all of the amino acids that are essential for growth and tissue repair. This is not the case with many vegetable proteins. Manioc, which contains almost as much protein as the other root crops, is very poor in certain of the essential amino acids and is thus by itself of only slight value as a protein source.

It is not necessary, however, that all the essential amino acids be present in a particular food. The total requirement may be derived from the proteins found in the variety of foods that are ingested together (Albanese and Orto, 1964:143f). Thus, the variety enjoyed by the Tsembaga not only spares them the gustatory boredom to

Table 9. Nutritive Value of Vegetable Portion of Tsembaga Diet Compared with FAO/WHO and New Guinea Food Standards

Consumers				Calories		Protein				Calcium	
Category	Number persons	Average weight (kgm)	Trophic units per capita	Rec. daily intake (Langley, corrected)[a]	Est. actual daily intake	Rec. daily intake (Langley, corrected)	Rec. daily intake (FAO/WHO, corrected)[b]	Est. daily intake (grams) min.[c]	Est. daily intake (grams) max.	Rec. daily intake (Langley, corrected) (mgms)	Est. daily intake (mgms)
Adult males	6	43.0	25.0	2,130	2,575	32	37	43.2	58.2	640	1,525
Adult females	4	38.0	21.0	1,735	2,163	32	33	36.3	48.9	640	1,281
Adolescent females	1	27.3	20.5	1,540	2,112	56	27	35.4	47.3	750	1,250
Children 5–10	2	18.6	13.0	1,157	1,334	44–53	21	22.5	30.3	890	793
Children 3–5	2	13.0	12.0	1,000	1,236	42	15	20.7	28.0	840	732
Children 0–3	1	7.7	8.5	800	875	33	10	14.7	19.8	940	519
Per capita average					2,015						

[a] Langley, 1947:134. Recommendations were made for people larger in stature than the Tsembaga. Venkatachalam (1962:10) has applied them directly to the Chimbu, among whom the average weight of adult males is 118 pounds. A correction was made for the difference between Tsembaga and Chimbu average weights.

[b] FAO, World Food Problems, #5, 1964, corrected for the biological value of vegetable protein, taken for the vegetable diet as a whole to be 70 (out of a possible 100).

[c] Minimum and maximum intakes are presented because of ranges in protein content attributed to various vegetables by various authorities. Authorities and values assigned by them appear in Appendix 9.

which some other New Guinea populations must be subject, but it also may improve utilization of the total intake. For instance, the amino acid contents of sweet potatoes are significant, but their utilization is limited by rather low proportions of methionine and cystine (Peters, 1958:40). Taros are good sources of methionine although poor in iso-leucine, in which sweet potatoes are well supplied (Peters, 1958:35). The presence of both items in the diet enhances the value of each, since they are frequently ingested at the same meal. Analyses of amino acid contents of many other Tsembaga vegetables are not available. It may be suggested, however, that the biological value (Albanese and Orto, 1964:140f) of the aggregate protein in the Tsembaga diet, although perhaps remaining low, is higher than would be the case if the diet consisted of fewer items, since it is reasonable to assume that the proteins in the many different foods ingested daily would complement each other to some extent.

There is a further problem, however. It may be noted in Table 9 that the intakes of adults meet the recommendations of the New Guinea Food Survey for protein, but the intakes of those in the younger categories do not. This may well be characteristic of people whose diets include root crop staples. To derive sufficient protein from root crops alone requires the ingestion of a greater quantity of tubers than children may be able to manage easily. While the caloric requirements of a five- to ten-year-old child may be fulfilled by two pounds of tubers, to meet the suggested protein allowances he would have to consume two to four times that quantity. The seriousness of this disparity is directly proportional to the extent to which root crops comprise the diet. Among the Chimbu, for instance, where roots comprise 82 percent of the total intake, five- to ten-year-olds are reported to ingest only 12.8 grams of protein a day (Venkatachalam, 1962:10). This is about half of the estimated Tsembaga intake, while the caloric intakes of the two peoples show nowhere near such a disparity.

The comparative advantage of the Tsembaga diet lies in the fact that starchy staples comprise only 65 percent of it. Furthermore, because the Tsembaga have available to them such protein-rich greens as hibiscus leaves, they need derive only 40 percent of their protein from tubers (see Appendix 8). The Tsembaga child's problem is thus not as serious as that of the Chimbu child, but it still may be difficult for him to obtain protein in a form sufficiently compact to be easily ingested. Animal protein is especially valuable to him not only because it is a more adequate protein but because it comes in a more concentrated form.

It must be kept in mind that the sample includes only five children and one adolescent, and the method used for apportioning the total weighed food sample among the various age and sex categories must have produced some error. A clinical assessment of the nutritive status of the children would have indicated more clearly than the comparison of estimated intakes to standard recommendations whether or not there was sufficient protein in their diets. It was not possible to arrange for such a clinical assessment but it may be noted that a number of the children had soft and discolored hair, frequently a sign of low or inadequate protein intake, and that the parotids of a few were slightly enlarged. No cases of either kwashiorkor or severe emaciation were seen, however, nor did any of the children seem especially dull. It is likely that their growth is retarded, but data are insufficient to assess this matter. Compared to American children Tsembaga children seemed small for their estimated ages. While this is to be expected among the children of a population in which adults average less than five feet in height, it is reasonable to believe that the small stature of the Tsembaga is at least partially attributable to low protein intake among children and, as will be discussed later, adolescents. It may be significant that the adults of the Fungai-Korama, a Maring group occupying a territory on the edge of a large expanse of virgin forest, are taller and heavier than the Tsembaga (Bick, personal communication). It is possible that their greater size is a result of larger amounts of game in their diet, but quantitative data are lacking.

That Tsembaga children do not seem to be suffering from more severe symptoms of protein deficiency, such as obvious kwashiorkor, suggests that Langley's recommendations are unrealistically high. Langley herself suggests care in their application, pointing out that they were derived from experimental work on Caucasian peoples and in addition include a "wide margin of safety" (1947:106). She made corrections for differences in stature and also on the basis of observations of "the mode of life and degree of activity of the New

Guinea native," but they were made without "physiological studies which would give a picture of the metabolism of Melanesian people" (1947:106).

More recent studies indicate that a considerably smaller protein ration may be sufficient even for children, and in column 8 of Table 9 the requirements set forth by the Joint FAO/WHO Expert Group on Protein Requirements (World Food Problems #5, 1964:6) are presented. The low incidence of severe pathology among Tsembaga children suggests that the FAO/WHO requirements may be more in accord with physiological facts than are Langley's recommendations. The FAO/WHO allowances do not, however, explicitly include a wide "margin of safety," and the fact that there is some pathology suggests that the Tsembaga achieve nitrogen balance at a low level. In view of this, the contexts in which the Tsembaga consume the limited amounts of animal protein available to them may be of considerable significance.

ANIMAL FOODS

A variety of domesticated and nondomesticated animal foods is available to the Tsembaga. Domesticated animals, it has already been mentioned, include pigs, cassowaries, and chickens. Nondomesticated animals include, in addition to wild pigs and cassowaries, marsupials, rats, snakes, lizards, eels, catfish, frogs, birds, bird eggs, bats, grubs, insects, and spiders. It might be mentioned that the Tsembaga are not and vehemently maintain their ancestors never were, cannibals.

No records were kept of the intake of nondomesticated animal foods, for a great part of the consumption, particularly of the smaller forms, takes place away from the houses. It is clear, nevertheless, that for certain categories of the population—married women and prepubescent children—small items such as rats, frogs, nestling birds, and insects form a part of the daily diet. The amounts are small, however, and probably do not contribute more than a gram or two to the daily protein intake. The larger items, eels, snakes, and marsupials, although less frequently taken, probably contribute more to the annual diet. Wild pigs, of which six were killed in 1963, contributed most of all the nondomesticated animal food to the diet in that year. Wild cassowaries are only infrequently obtained by the Tsembaga; none were either shot or trapped in 1963. These birds may be more important in the diet of other Maring groups.

The apportionment of animal foods throughout the population is not even. Taboos against eating certain animals burden various categories of persons. The most extensive of these are suffered by *bamp kunda yu* (fight magic men), those who perform the rituals associated with warfare. The taboos applicable to other men who have participated in fighting are slightly less extensive. For both of these groups the consumption of many kinds of marsupial and all snakes, eels, catfish, lizards, and frogs is proscribed. The ideological basis for the restriction on marsupials is not clear. It may be mentioned, however, that all of the proscribed varieties live in the high-altitude virgin forest and are associated with the red spirits, with whom the men suffering the proscription have a special relationship. The reptiles, amphibia, and fish are not eaten because their "coldness" is regarded to be inimical to the "hotness" of ritual knowledge and ritual experience. In addition to these supernaturally sanctioned taboos, men also simply avoid certain animal foods. Any man, other than fight magic men, may eat rats, but no man would. They are said to be small things fit only for women and children. The same is said of small birds, bats, and most insects.

Adolescent boys suffer a somewhat different set of taboos and avoidances. The ideological basis for many of them rests upon the effect that eating certain foods might have on the massive coiffure, called a *mamp ku,* which a boy allows to grow shortly after secondary sexual characteristics appear. If a marsupial with loose fur is eaten, for instance, it is said that the hair will become loose. If wild pig is eaten, the hair will be filled with lice, like that of the wild pig.

The restrictions to which adolescent girls and women of childbearing age who have not yet borne children are subject are more limited than those that apply to adolescent boys. They do, however, pertain to the two most important wild animal foods: wild pig, the consumption of which, they believe, will give them lice, and rats, which will impart a bad odor to the vagina, making them unacceptable to men.

Children and women who have borne children are, as categories, subject to no taboos whatsoever. Thus, when a man or adolescent boy kills a snake, lizard, small bird, rat, or marsupial, he gives it to a woman or child.

The taboos operate, in short, to direct most of the subsidiary sources of animal protein to two categories very much in need of them: women and children. The deprivation of adult males, who as an ontogenetic category need the least protein, in favor of women and children may be of some advantage to the population. On the other hand, the deprivation of adolescents in favor of women and children is not as clearly advantageous. Protein requirements during adolescence are high, for this is a period of rapid growth and development. It may be observed that while some of the other possible effects of low protein intake during adolescence (particularly mental dullness, which was not observed among Tsembaga adolescents) are clearly injurious, the inhibition of growth and the retardation of sexual development (Hipsley and Kirk, 1965:14) may not be disadvantageous. The advantages conferred by an additional inch or two and an additional few pounds could not be great, and the delay of menarche for a year or two might even be of advantage to the population as a whole. At any rate, while deprivation of protein at this time may result in suppression of the individual's growth, this may not be a high price to pay for more adequate protein for mothers and small children.

Since the Tsembaga, unlike some other Marings, keep almost no cassowaries and since chickens are few, pigs are the major source of food from domesticated animals. During the fieldwork period a pig festival was in progress, and the usual patterns of pork consumption were obscured by the special demands of various ritual and ceremonial events. Therefore, although it was possible to collect information from informants concerning pork consumption in nonfestival years, quantitative data are lacking.

It should be mentioned first that religious prohibitions suffered by men are not restricted to wild animal foods. During the greater part of the ritual cycle they may not eat the flesh of pigs killed in connection with the festivals of other local populations. Women, children, and adolescents, however, may eat such pork and are thus the full nutritional beneficiaries of these presentations.

Personal preferences also may play a nutritionally significant role in the apportionment of pork. Men receive larger portions of the cuts they deem most desirable. These are the fatter parts of the animal, with the belly being most appreciated. Women, children, and adolescents on the other hand receive larger portions of the lean. In short, the protein intakes of women and children are benefited by preference as well as taboo in the matter of domestic pork.

Pigs are seldom eaten on occasions that are not ceremonial. They sometimes become sick and die, of course, and an animal that has succumbed to disease will be eaten unless its carcass has putrefied before it is recovered. Some also die of arrow wounds, for pigs are not infrequently guilty of invading gardens and are shot by garden owners. No Tsembaga pigs were killed for this reason in 1963 because, informants said, almost all of the gardens were separated from the residences by a fast-rushing and steeply banked stream. But news reached me of six animals being shot among the neighboring Tuguma during the same year. Feelings run high after such killings; the owner of a slain pig is likely to make an attempt on the life of the garden owner, his wife, or one of his pigs. If a peaceful settlement can be achieved the flesh of the slain animal is apportioned among the agnates of the two principals to the dispute, who may redistribute it to others. The owner of the slain pig is then supposed to repair the fence that the animal damaged in breaking into the garden, and the garden owner is supposed to present him with a piglet to replace the one killed. When this animal matures, it is killed and both parties to the dispute share its flesh evenly, according to informants.

Pigs are deliberately killed by their masters during nonfestival years only for the ceremonial fulfillment of obligations or in the context of rituals associated with misfortune and emergency.

The nonfestival obligations requiring the slaughter of pigs are mainly those associated with marriage.[3] Prestations of valuables are made to affines from time to time in connection with the marriage itself, with the birth of children, and with the deaths of either of the marriage partners or their children. These prestations are almost always accompanied by pork, but do not always follow immediately upon the assumption of the obligation. Those made in connection

3. Young men occasionally kill pigs and present the flesh (some of which is salted) to older men from whom they wish to obtain knowledge of important magical procedures. Such transactions are rare. Some men are never involved, while most others enter into them only once or twice in their lives.

with the birth of children are often deferred until the next pig festival, and such delays seem often to occur even in the case of marriage prestations.

The demands arising from misfortune and emergency probably result among the Tsembaga in the greatest amount of pig slaughter and consumption during years unmarked by pig festivals. (This may not be the case with all Maring. Cherry Vayda in a personal communication reports that among the Kauwasi, a Maring group living in the Jimi Valley, more pigs are killed in nonfestival years for affinal prestations.[4]) Most frequently the misfortunes calling for pig slaughter are illness and injury, but during the period of fieldwork the Tsembaga, because they were conducting a pig festival, did not kill pigs immediately upon injury or during the course of illness. Instead, they promised the ancestors that they would give them animals at the termination of the *kaiko,* when they were planning to kill them anyhow. But all informants agree, and some information from other local populations indicates, that in nonfestival years pigs are sacrificed soon after a person is injured or during the course of an illness that hasn't responded to less radical treatment (magical procedures), and if illness is protracted a number of pigs may be killed.

Death as well as illness and injury demands the slaughter of pigs. If the deceased was a married woman the slaughter is likely to be

4. Data are insufficient, but I believe that this difference may be a function of differences in horticultural practices. Clarke's (1966:350) observations indicate that sweet potatoes are relatively more important among the Kauwasi than the Tsembaga. This may be in part attributable to the somewhat higher altitudes at which the Kauwasi are gardening (up to and a little beyond 6,000 feet), and perhaps to changes in soil structure on Kauwasi territory, which seems to have been occupied for a longer period than Tsembaga territory. At any rate, fallows are shorter on Kauwasi garden sites and there are clear signs of some degradation of Kauwasi land in general: about 20 percent of the territory is under stabilized short grass (Clarke, 1966: 350ff).

The Kauwasi seem to keep more pigs than do the Tsembaga (no censuses have been taken), and the greater importance of pigs is probably not unrelated to the greater importance of sweet potatoes. When one informant was asked why the Tsembaga kept fewer pigs than the Kauwasi, he responded in effect that it is because Tsembaga soil is softer and more fit for taros and yams. Kauwasi soil is harder and more suitable for sweet potato. Since the Kauwasi must grow more sweet potatoes they keep more pigs, which they eat more frequently. It may be that we can see here a long-term succession in horticultural practice. In recently entered areas there is greater emphasis on taro and yam cultivation. With continued occupation there are structural changes in the soil that lead to a greater emphasis on sweet potatoes and pig husbandry. For an excellent discussion of possible successions see Clarke (1966).

of greater magnitude than for other categories of persons. Instances are reported in which men have killed all of their deceased wife's pigs, for there were no other women to look after them. As in the case of illness, the Tsembaga in 1963 killed no pigs at times of death. They merely promised animals at the termination of the pig festival.

The slaughter of pigs is also demanded by the rituals associated with warfare. The number of pigs killed is directly related to the duration and severity of the fighting. This matter will be treated in more detail in the next chapter.

The occasion that defines when a pig is to be killed also designates who is to eat it. Pigs sacrificed during the rituals associated with the actual prosecution of warfare are consumed only by the men participating in the fighting. When pigs are killed in connection with illness, injury, or death, the pork liver is served to the victim, if he is alive. If the sacrificial pig belongs to the victim or his agnates, informants told me, the remainder of the flesh could be consumed only by the agnates or, in the case of a married woman, by her husband's agnates as well. It was said that the sacrifice is made to spirits of deceased ancestors (to the red spirits if the upper part of the body is affected, to the spirits of rot if suffering is in the lower portion of the body) to mollify them if they (or one of them) have visited the affliction, or to strengthen them so that they might assist in recovery if the victim has been the target of a hostile spirit, sorcerer (*kum yu*), or witch (*koimp*). The recipient spirits, informants said, would be angered if the flesh of animals sacrificed to them under these circumstances were to be distributed to nonagnates and they would either withhold their assistance from the victim or continue their attack upon him, with the result that he would die, and they would also afflict illness upon the nonagnatic consumers. Despite these statements, personal communications from Cherry Vayda and Georgeda Bick, who returned to the Maring area in 1966, indicate that resident nonagnates may also receive portions of the sacrificial flesh. It is sometimes the case, moreover, that a cognate or affine contributes a pig to be sacrificed for the benefit of a victim of illness or injury and in such instances a larger but still restricted number of people may partake of the pork.[5] (Such contributions by

5. I did not perceive the full significance of pork consumption in the context of misfortune while I was in the field, and therefore did not collect information on this

nonagnates are expected to be reciprocated at later pig festivals or rewarded in some other way. I know of one instance in which the victim of illness, having recovered, transferred to his nonagnatic benefactor perpetual rights in a number of garden sites.)

In sum, pigs killed in the context of misfortune or emergency apparently tend to be consumed by those who are either victims of or participants in the event, or by those close to them. It is interesting to note, in contrast, that there is a tendency to distribute widely the meat that results from the killing of pigs for reasons other than misfortune or emergency. Pork that is received in bride or child payments, for instance, is consumed not only by members of the recipient exogamous group; some of it is invariably redistributed by them to coresidents and affines. Pork resulting from the death of pigs from accident or natural causes is also widely distributed. It may be mentioned that the consumption of the flesh of wild pigs is also not restricted to the agnatic group of the slayer. There seems to be, indeed, a tendency to distribute the flesh of these animals as widely as possible. At least all of those who participate in the hunt share in the flesh, which is apportioned by the slayer. If the animal is large, others do, too. In the case of the two largest wild pigs killed in 1963 every Tsembaga received a portion.

It is a matter of physiological interest that most pork consumption in years in which pig festivals do not occur probably takes place in contexts of misfortune or emergency, and that the distribution of meat in these instances is restricted. It is reasonable to assume that misfortune and emergency induce in those experiencing them a "stress reaction," a complex of physiological changes that result directly from or in response to the emergency with which the organisms are confronted. In the case of the sick and injured, this is obvious. It is also the case, however, that stress reactions occur among people who are experiencing rage or fear (Houssay, 1955:1096) or even prolonged anxiety (NRC #1100, 1963:53).

One aspect of stress reactions is "negative nitrogen balance" (Houssay, 1955:440), that is, a net loss of nitrogen from the body. This may not, if the stress period is not unduly prolonged, be a

---

subject in the detail that is warranted. Georgeda Bick and Cherry Vayda will publish elsewhere the details of apportionment and context definition for pig sacrifice, based upon their research into this subject in 1966 and 1967.

serious matter for organisms that have previously achieved nitrogen balance at a high level. Negative nitrogen balance, according to Moore (1959:439), will not, for instance, impair wound healing in a normally well-nourished patient.

However, organisms that are in nitrogen balance at a low level previous to their exposure to stress may experience difficulties. As long ago as 1919 it was reported that the healing of experimental wounds in animals on low protein intakes was slower than healing in animals on high-protein diets (Large and Johnston, 1948:352). In a more recent discussion Moore also notes differences in healing between poorly and well-nourished animals and surgical patients, and states that "it is appealing to explain this . . . in the labilization or translocation theory; the well nourished animal has more body protein available for labilization or translocation [from undamaged tissues to damaged structures]" (1959:102). Zintel (1964:1043) has also pointed out that "hypoproteinemia predisposes to poor wound healing or disruption of wounds, delayed healing of fractures, including anemia, failure of gastro-intestinal stomas to function, embarrassment of pulmonary and cardiac function and reduced resistance to infection." Lund and Levinson, in a discussion of surgical patients, noted yet other ill effects of protein depletion upon the injured some years ago. These included decreased resistance to shock, hypotension, lower basal metabolism rates, polyuria, lack of appetite, weakness, and mental changes including confusion, lethargy, and depression (1948:349). Lund and Levinson suggest a "great increase" in the protein intake of individuals traumatized by surgery (1948:350), and their view is shared by Elman (1951:85, 100). The more recent writers do not necessarily suggest such a course for individuals who were well nourished previous to traumatization, but Zintel (1964:1043ff) underlines the necessity to alleviate protein deficiency, if it should be present or should develop in surgical patients.

It is not only victims of trauma who may be put at a disadvantage by poor protein nutrition. In febrile diseases "the underfed febrile patient is thrown into a negative nitrogen balance, which is far from desirable since it predisposes to hypoproteinemia and anemia, interferes with the body's anabolic defense mechanisms, and delays convalescence" (Burton, 1959:230). A protein allowance somewhat in excess of usual requirements may therefore be indicated for fever patients (Burton, 1959:231).

It may be the case that protein intake not only affects the prognosis for ill or injured organisms, but may also have a bearing upon susceptibility to some diseases. While the subject is not well known, Berg (1948:309ff) has remarked that "numerous incidental observations have indicated the possibility that proteins play an important part in immunity," and relates this to their role in the production of antibodies and the part they probably play in phagocytosis as well. It may be of importance here to note that the status of a protein-depleted organism may be significantly improved by the intake of high-quality proteins for relatively short periods of time. Berg (1948:311) states that experiments indicate that repletion of depleted animals "by feeding a high quality protein for as little as two days before antigenic stimulation leads to a detectable increase in antibody production; in only seven days the capacity [of the previously depleted animals] approximates that of the control animals [which had previously been fed an "adequate diet"]." In a more recent discussion Axelrod (1964:654) states that "the detrimental effects of specific dietary deficiencies upon the development of acquired immunity in experimental animals have been amply documented. In particular, the requirements for amino acids . . . are recognized." He warns, however, that "the significance of animal experimentation must . . . be evaluated critically," and remarks that "the relationship of nutritional state to acquired immunity in man remains indeterminate" (1964:655). Zucker and Zucker, in a discussion of the relationship between nutrition and the resistance of uninfected animals to infection, make a similar observation concerning the role of proteins: "The available data are largely inconclusive or uncertain as to interpretation" (1964:643).

It appears, then, that illness and injury are marked by negative nitrogen balance, which can have dangerous implications for protein-depleted organisms. The effects of negative nitrogen balance, however, may be offset rather quickly by the ingestion of large amounts of protein. It also may be, although this is far less certain, that the consumption of high-quality protein for relatively short periods of time may significantly improve the ability of uninfected organisms to withstand infection. I suggest that, given the adequacy of the protein derived from vegetables and nondomesticated animals for maintaining the Tsembaga in nitrogen balance at low levels in the absence of stress, the practice of sacrifice in situations of misfortune and emergency is a highly effective way to utilize the scarce and costly pigs. Individuals who are already traumatized or diseased are provided with high-quality protein, which may go far to offset the nitrogen losses they are already experiencing as a direct result of the injuries to their bodies, and which also assists them in producing sufficient antibodies to resist infection. Those close to the victims also receive protein, which not only may offset the nitrogen loss resulting from the anxiety they may be experiencing but also might possibly prepare their bodies better to withstand the injuries or infections likely to be forthcoming if the victim is suffering from a contagious disease, for example, or was wounded in warfare that must be continued.

In sum, while the protein content of the everyday diet of the Tsembaga is probably adequate for everyday activities, it may be less than adequate in stress situations. The practice of killing and consuming pigs in connection with emergency and misfortune tends to provide physiological reinforcement when it is needed to those who need it. The contribution of pork to the Tsembaga diet thus seems to be of much greater importance than is indicated by the amounts actually consumed. Clearly this suggestion has implications for future anthropological analyses of animal sacrifice, and appropriate physiological tests should be undertaken wherever possible; until they are this formulation must remain hypothetical only. It may be said nevertheless that there are strong reasons to believe that the ritual regulation of pork consumption by the Tsembaga makes an important contribution to a diet that maintains the population in adequate health at a high level of activity.

THE LIMITS OF THE SYSTEM

An attempt has been made here to specify the trophic requirements of the Tsembaga, the procedures by which they fill them, and the effects that these procedures have upon the environment in which they take place. Now a computation will be attempted of the carrying capacity of Tsembaga territory, that is, the maximum number of people and pigs that could be supported on it for a period of indefinite length, through the procedures described above, without any change in the intakes of individual Tsembaga and without depletion of the environment. These estimates must be made before we can examine the role of the ritual cycle in regulating the relations of the Tsembaga with the nonhuman components of their immediate, or territorial, environment.

It should be emphasized that carrying capacity figures presented here do not necessarily constitute approximations of the number of people who could survive on Tsembaga territory. Estimates of carrying capacity, as the term is used in anthropology,[6] do not necessarily provide such information. They are, rather, approximations of the numbers of organisms that may be supported within a given area *without inducing degradation of the environment.* In fact, populations may, and often are, limited by environmental factors that become significant below carrying capacity in this sense. Density-dependent epidemics and predators may be cited as examples. It should also be kept in mind that Liebig's law of the minimum and Shelford's law of tolerance can come into operation at levels below those sufficient to degrade the environment. A carrying capacity figure indicates the numbers of a species that can survive within a designated area only if ranges set by other factors have not been exceeded at lower population levels and the population is limited by its gross food supply, that is, by the production of usable biomass in the area it exploits.

It is impossible to say whether or not the probable limits of Tsembaga population are set by the productivity of their territory or by other factors. Nevertheless, a carrying capacity estimation allows us to approximate the levels at which variables in the Tsembaga ecosystem become destructive of that system.

Difficulties Concerning Carrying Capacity

The advantages that may be gained from a calculation of carrying capacity should not blind us to the difficulties, both conceptual and methodological, hidden beneath the simplicity of the concept. Some

6. In animal ecology the term *carrying capacity* is sometimes used in a slightly different sense: to refer to the maximum number of animals of a particular kind that can survive in an area (Andrewartha, 1961:154). This number is a function of the processes that actually limit the size of the population. These factors do not necessarily include environmental degradation.

of the problems encountered in Tsembaga data are discussed in Appendix 10, but it is useful to call attention to some of the more general ones here.

1. Both conceptual and methodological problems are presented by the concept of environmental degradation. If it is taken to mean reduction in the productivity (measured by the amount of biomass produced within a specified area during a specified time period) or decrease in the organization (complexity and orderliness) of the ecological system, then the mere replacement of a climax community by a community based upon cultivated plants is likely to represent environmental degradation. Cultivated communities are in most cases simpler in structure than the climax communities they replace, and there is strong reason to believe that their productivity is usually lower (Allee et al., 1949:478 passim, 507; Odum, 1959:76). Climatic or edaphic climax communities represent optimal adaptations to their abiotic habitats, and it is only through the alteration of the abiotic habitat (for example, through irrigation or the application of fertilizer) that communities dominated by man are likely to exceed them in productivity.

It is furthermore the case that the cultivated community will, almost without exception, be less stable than the climax community. The climax community together with its habitat constitutes a self-regulating system. As such, it will endure through long periods of time unless it is subjected to such agencies as climatic change, cataclysm, or the intervention of certain organisms such as agricultural man or blight parasites. The cultivated community, on the other hand, is not per se a self-regulating system. Many of the plant species that it includes may not have the ability to propagate themselves and they require protection from competitors, such as weeds and second-growth trees, which are often as well or even better adapted than they are to the local conditions of the garden and field or the general conditions of the region. The maintenance of the cultivated community, like any community, depends upon the constant functioning of its dominant species. Man, however, is less reliable in such a role than the "A" stratum trees that dominate the rain forest. But whether or not cultivation degrades the environment in an absolute sense, it must be remembered that, since the Neolithic, man has not been primarily a member of climax associations. He is, rather, a form that thrives mainly in the highly modified ecosystems that he himself dominates. It is not of direct concern to him that the productivity of such ecosystems is lower than those of the climatic communities they replace. The amount of usable biomass with which they provide him is much higher.

The instability of the cultivated ecosystem is more directly relevant to man. On the one hand there is the possibility of the reassertion of the succession leading to the climatic or edaphic climax characteristic of the region. While this may not be regarded as degradation in the absolute sense, such a process results, at least temporarily, in less land available for cultivation. It also is likely that considerable labor must be expended to reclaim such areas for cultivation. On the other hand, and more likely, there is the possibility of further absolute degradation leading to the establishment of a "disclimax," such as the anthropogenic grasslands found in some parts of New Guinea where forests and gardens once flourished. While gardens made in such grasslands yield well, they do so only if special techniques, which are costly in terms of labor, are employed.

In light of these considerations, in a discussion of the environment of human groups we should distinguish between "absolute degradation" and "anthropocentric degradation," the latter term referring to any process that lowers the productivity of biomass useful to man per unit of area or labor input or both. To put it in slightly different terms, it refers to deviation from conditions that are optimal for the survival of a human population dependent upon a particular set of resources and equipped with a particular set of means for acquiring them.

The question of whether or not degradation, absolute or anthropocentric, is widespread over Tsembaga territory was touched upon earlier in the discussion of the adequacy of fallow periods. Evidence indicated that except in highly localized areas there is no degradation of either sort. The assumption will be made for the purpose of estimating carrying capacity that degradation is not occurring, but it should be reiterated that the evidence falls short of demonstration. As Street has underlined in a recent article (1965), in many instances it is virtually impossible to assess environmental degradation, particularly in the structure and content of the soil: "Deterioration of the land is a cumulative process, and short term processes may be so slight as to be exceeded by errors in measurement."

2. A related problem concerns the lengths of fallow periods when carrying capacity is estimated for people depending upon swidden gardens. Shifting horticulture is frequently practiced by people who are unaccustomed to keeping careful track of the years, and the length of fallow periods may vary from site to site. It often becomes necessary to estimate fallow periods through involved means that may produce inaccurate results. Moreover, even if the estimation is accurate the actual fallow periods may be longer or shorter than they need to be, and their use in carrying capacity computation may produce inaccuracies.

3. Carrying capacity estimates, including those to follow, assume not only that the inventory of crops remains unchanged, but that the proportions in which these crops are planted will also remain constant. People practicing mixed crop, shifting horticulture (and the Tsembaga are no exception) readily accept new crops and new varieties of old crops, as Street (1965) has pointed out. To the extent that new introductions or changes in planting proportions will produce changes in yields per unit area, the assumption of a constant inventory of crops and constant planting proportions will deflect an estimate of carrying capacity from true values.

4. Carrying capacity estimations include as one variable the area of cultivated land required to provide a population with food. The value for this variable is usually derived from the practice of the people under study. It has already been pointed out that an areal figure alone does not indicate the nutritional level at which the population is being supported, or whether this level is adequate. An estimation was made above of the intake of the population and the adequacy of this intake. There remains, however, an associated problem. Except in rare cases local populations of human beings are made up of individuals in a number of age and sex categories, each of which has different trophic requirements. Estimations of carrying capacity are based upon the assumption that the proportion of individuals in each age-sex category will remain constant. This assumption underlies the computations that will follow, but no doubt it differs from reality to some extent. Among all human populations, and especially among small groups living under primitive conditions,

the proportions of individuals in the various categories is constantly shifting.

The Carrying Capacity of Tsembaga Territory

A number of formulae have been proposed for applying the concept of carrying capacity to human populations. Those of Allen (1949:74), Brookfield and Brown (1963:108f), Carneiro (1960:230–31), Freeman (1955:133), and Loeffler (1960:41) are addressed particularly to the problems inherent in arriving at carrying capacity for groups practicing shifting cultivation. Among the variables considered are amounts of land put into cultivation, duration of harvesting, length of fallow, and total area of arable land. Brookfield has further categorized garden land into various types and takes into consideration grazing land.

The formulae proposed are generally simple ones. Carneiro's, which follows, has been used here:

$$P = \frac{\frac{T}{(R+Y)} \times Y}{A}$$

Where:

P = The population that can be supported.
T = Total arable land.
R = Length of fallow in years.
Y = Length of cropping period in years.
A = The area of cultivated land required to provide an "average individual" with the amount of food that he ordinarily derives from cultivated plants per year.

Through the application of this formula and subsidiary procedures outlined in Appendix 10, a number of carrying capacity estimates have been derived.

Two values have been used for variable T, total arable land: (1) land that at the time of field study was either in cultivation or under secondary forest, and (2) land deemed arable but under high forest. Because these latter areas have not been cultivated in the recent past, because they are important sources of nondomesticated resources, and because, being at high altitudes and frequently covered by cloud they would doubtless be only marginally productive under cultivation, it was considered advisable to segregate them.

Because pigs have trophic requirements of a magnitude comparable to those of humans, the pig population had to be taken into consideration in estimates of carrying capacity for humans. Estimates were therefore made of carrying capacity for humans with the pig population at its minimum and maximum sizes. Minimum pig population size was assumed to be represented by the number of pigs surviving the pig festival in 1963. At that time the ratio of pigs, averaging 60 to 75 pounds, to people was .29:1. Maximum pig population size was assumed to be represented by the number of animals the Tsembaga had at the beginning of the pig festival in 1962. The ratio of pigs, averaging 120 to 150 pounds each, to people at that time was .83:1.

Because all Tsembaga gardens made in 1962 were measured by chain and compass, only one calculation was made for human carrying capacity with maximum pig population. Only some of the Tsembaga gardens made in 1963 could be measured, however, and it was necessary to calculate the value of variable A, the amount of land put under cultivation per capita. Three methods were used in this calculation. The results produced by using the extreme values for variable A are reflected in the range shown for human carrying capacity with the pig population at minimum size. It should be noted here, however, that two of the three calculations for variable A were separated by a difference of only 4 percent while the third calculation was separated from one of these by 27 percent. The use of the aberrant value for variable A produces the lower figure, 290 persons, for carrying capacity with the pig population at minimum. For reasons discussed in Appendix 10 this figure is deemed to be less reliable than the higher figures.

Carrying capacity for pigs, holding the size and composition of the human population constant, was also estimated. Two methods were used, and in applying one of these methods the two extreme values were again used for variable A. The lower figure, 142 pigs, was produced by the use of the aberrant value for variable A and is deemed to be less reliable than the higher figure, 240 pigs (of 120- to 150-pound size). The results of the various estimates are summarized in Table 10.

Table 10. Estimates of Tsembaga Territory's Carrying Capacity, Human and Pig Populations

	Secondary forest	High forest	Total
Human population (pig population held to minimum)	290–397	44–60	334–457
Human population (pig population maintained at maximum)	251	38	284
Human population, mean carrying capacity	271–324	42–49	313–373
Pig population (human population held constant)	142–240	62	204–302

Several comments need to be made concerning the figures in this table. It should be repeated that the figure of 290 persons, if the pig population is never allowed to exceed minimum size, is probably too low. On the other hand, the figure of 397 persons may be too high for the number of people that can be supported by land under secondary forest. A calculation employing the intermediate value derived for variable A produces a carrying capacity figure of 383 persons. Averaging the three calculations yields a figure of 356 persons.

It should be emphasized here that the carrying capacities for humans with the pig population at minimum represents an estimate of the number of people that could be supported on Tsembaga territory if the ratio of pigs (of 60- to 75-pound size) to people was *never* allowed to exceed .24:1. Similarly, the figures for carrying capacity for humans, pig population at maximum, is an attempt to estimate the number of people who could be maintained if the ratio of pigs (of 120- to 150-pound size) to people was *constantly* maintained at .83:1. In fact, the pig population fluctuates, and the actual carrying capacity for humans lies between the figures estimated in connection with the extreme sizes of pig herds. Where in this range the actual carrying capacity will fall depends upon demographic processes within the pig population, which itself is dependent upon so many factors, including the changes in the human population, that precise estimation would be impossible. No more could be done here than to strike means between the estimates of carrying capacity with minimum and maximum herd sizes. These have been represented as mean carrying capacity.

Figures for carrying capacities for pigs must be interpreted in a similar way. These figures represent attempts at estimating the number of pigs of 120- to 150-pound size that could be supported *continuously* on Tsembaga territory, if the size and composition of the human population remained constant. Since the size of the pig population fluctuates it could doubtless exceed considerably, for years at a time, the figures presented here without degrading the territory. Thus, although the Tsembaga pig population stood in 1962 above one of the figures derived for carrying capacity this does not mean, granting the correctness of the figure, that the number of pigs had actually exceeded the capacity of the territory to support pigs.

If the estimates presented here are even approximately accurate, the Tsembaga were well below the carrying capacity of their territory in 1962 and 1963. There have been times in the past, however, during which this might not have been the case. In the last chapter it was suggested that a population of 250 to 300 persons is indicated for the period thirty to forty years prior to fieldwork, and at that time the Tsembaga had less land. It was during that period that they annexed about 135 acres (surface area) of arable land by driving out the Dimbagai-Yimyagai, who had previously lived immediately to their east.

It may be that differences in horticultural and pig-herding practices compensated for the larger populations in earlier times. Information concerning this is fragmentary and difficult to quantify. People do say that the number of pigs regarded as sufficient for a pig festival was smaller in earlier times. Considering the amount of labor necessary to make gardens with stone tools, this would not be surprising. People also say that they previously ate more bananas, and it may be that before the arrival of steel tools smaller gardens were kept in production longer, with third- or even fourth-generation bananas making a much more important contribution to the diet than they did in 1962–63; however, information is conflicting and unclear.

It may well be asked why additional virgin forest was not put into cultivation during periods of high population. Use of these lands could have substantially increased the number of people who could be supported on Tsembaga territory. That they were not utilized, at least in fairly recent times, is not sufficiently explained by the fact that cutting gardens is ritually proscribed in much of the area. Other factors may be suggested, however.

First, the Jimi Valley lands are both distant and exposed to enemies. The latter reason was agreed upon by all informants, who said that their ancestors did make some gardens there, but the area has been avoided since the development of the enmity with the Kundagai. Even in the absence of such an enmity, however, the regular utilization of lands three or four hours' hard walk (including climbs of 2,000 to 4,000 feet in both directions) from other arable lands could not be very attractive.

Secondly, these are marginal lands. Considering both the low yields of high-altitude gardens and the long fallow periods they seem to require, these lands are judged to be less than 40 percent as productive as the best land that the Tsembaga have under production. Clearing virgin forest without steel tools must have been extremely arduous, and it may have been that people would have eaten less, kept fewer pigs, lived or gardened uxorilocally or sororilocally, or seized their neighbors' land rather than expend effort in cultivating such poor ground.

Third, it should be kept in mind that these areas in their condition as virgin forest are important sources of animal food, fiber, building materials, and other items. Data are insufficient to estimate the extent to which the deprivation of these items, resulting from removal of the present vegetation cover, would adversely affect the Tsembaga. It is surely the case, however, that although the estimations of carrying capacity here have been based upon domesticated foodstuffs, other items are necessary to the well-being and even survival of the population.

SYSTEM AND ECOSYSTEM

In this discussion of the Tsembaga and their environment, the Tsembaga have been taken to be an ecological population in an ecosystem bounded by the limits of what is recognized as their territory. My purpose has been to describe and, where possible, assign values to components of this system, rather than to analyze its functioning.

The designation of dimensions of the phenomena described above as variables in a system has an advantage beyond that of facilitating the expression of their interrelations. It also makes it possible to specify approximately the limits within which the system—that is, a particular set of dynamic interrelationships among specified variables—can continue to exist.

First, the parameters of the system may be designated. These are abstracted from conditions that affect variables within the system but which vary independently of them. Included among the parameters of the system that is emerging here are aspects of such factors as terrain, altitude, temperature, rainfall, and insolation.

The presence or absence of variables within a system cannot be explained in terms of the other variables in the system. Presence or absence can, however, sometimes be explained in terms of the parameters of the system. Brookfield's (1964) discussion of insolation, sweet potato cultivation, and population density in the New Guinea Highlands is a recent example of the explanation of presence or absence by reference to the parameters of systems. Kroeber's (1939) discussion of maize is another example. The specification of parameters is, of course, especially important in comparative studies. A description of the ways in which the Tsembaga relate to their immediate environment may be adequate without reference to rainfall, temperature, or altitude. The same is the case for the Chimbu people. In comparing Tsembaga and Chimbu subsistence procedures, diet, and nutritive status, however, we would note important differences. These are in large measure due to parameters of the system: for instance, the limits of toleration of certain plants, notably many protein-rich greens, probably exclude them from the Chimbu crop inventory, and therefore from the Chimbu diet, from which they are apparently absent.

In addition to designating limiting conditions or parameters, it may also become possible, when dimensions of phenomena are regarded as variables in a system, to discover the "system-destroying" levels of these variables.

The formulae for computing carrying capacity may be regarded as summary statements of the interrelations of the variables that they include. They do not, however, elucidate the mechanisms by which the values of the variables are regulated. A carrying capacity figure indicates ranges of values within which variables must remain if the system is to endure, but it does not indicate how values are kept within those ranges. In the course of the descriptions in this chapter, some processes and mechanisms that contribute to maintaining variables with "ranges of viability" have been mentioned; for example, the effect of selective weeding upon the length of time required for the fallow on garden sites was suggested. In a later chapter the ritual means by which other variables are kept from reaching system-destroying levels will be discussed.

CHAPTER 4

Relations with Other Local Populations

Just as the Tsembaga form part of a network of relationships with nonhuman components of their immediate environment, so do they participate in relationships with other local populations similar to themselves living outside their territory.

THE LOCATION OF OTHER GROUPS

The population size and territorial area of the Tsembaga fall around the middle of the ranges of Simbai Valley Maring groups. The small size of territories and the relatively large numbers of people who occupy them result in the proximity of both the residences and subsistence activities of adjacent groups. The houses of one of the clans belonging to the Tuguma local population immediately to the east are not more than a half mile from the houses of the Tsembaga. The residences of the Kundagai, the local group to the west of the Tsembaga, are about one and one half miles away. Both the eastern and western borders follow certain watercourses from near the top of the Bismarck Mountain to the Simbai River.

The Tsembaga refer to the Tuguma, along with themselves, as "*amaŋ yindok*" ("inside" or "between" people), meaning that they and the Tuguma form an enclave of friendly people occupying a continuous area bordered by hostile groups. The territory to the east of the Tuguma is occupied by their enemies of long standing, the Kanump-Kaur, and the Tsembaga fought four wars against their western neighbors, the Kundagai, during the forty or fifty years before 1962.

This spatial distribution of friends and enemies seems to be typical of the Maring. According to Vayda, almost every Maring local population shares at least one border with an enemy, and in almost every case the antagonists are located on the same side of the valley. Enmities between groups separated by the major rivers in the valley bottoms, the Simbai and the Jimi, do occur, but they are rare and short-lived. Enmities between groups separated by the mountain ridge do not, according to Vayda, occur at all.[1] It is no doubt relevant that the trade routes generally run across the grain of the land. Commodities that are not provided in a group's own territory, or are provided only in insufficient quantities, are obtained from people living across the river or over the ridge in the next valley.

It may also be relevant that a man would find it difficult to exploit land separated from his residence by a major ridge or river. Because residences are in the middle altitudes, the walk to land across the river or in the next valley is long, and a climb of several thousand feet is inevitable. During some parts of the year, moreover, the major rivers become impassable for days or weeks at a time. Only by moving his residence can a man conveniently utilize such distant ground for regular root crop gardens, and this alternative is not likely to be attractive to most people, even when they are short of land.

It is not surprising, then, that friendly relations generally prevail between groups separated by mountain ridge or river. Their relationship is mutually advantageous in that they supply each other with needed or desired commodities, and they are noncompetitive because their lands are not of great use to each other.

Conversely, it is not surprising that relationships between groups living side by side on the same valley wall are frequently antagonistic. Such adjacent groups do not depend upon each other for commodities. They are not bound together by the material exchanges that characterize the relationships of people whose territories are separated by mountain ridge or river. Their proximity, furthermore, increases the probability of friction between them. While it is extremely unlikely, for instance, that the pigs of groups separated by mountain or river will damage each other's gardens, it is a common occurrence among groups living adjacent to each other on the same valley wall.

1. The term *enmity* refers here to the relationship between groups who have opposed each other as principal combatants (see p. 117) in rounds of warfare.

It is also the case that some of the garden land of the neighboring group is likely to be as close to a man's residence as some of his own. If a group were short of land its members would look to the conveniently located land of the adjacent group rather than to land across the river or the mountain.

In contrast then to relations between groups located across the grain of the land from each other, the relations of groups occupying adjacent territories on the same valley wall may become competitive in respect to land and are not mutually beneficial in exchanging those goods in which the territories of each might be lacking.

FRIENDLY RELATIONS

Relations with friendly groups are concerned with and bound by exchanges of women and goods and may therefore be discussed in terms of marriage and trade.

Marriage

Detailed information concerning Maring marriage will be presented elsewhere; a brief summary is sufficient for the purposes of this study. It has already been mentioned that the Tsembaga state a preference for wives who are also Tsembaga by birth, and that this preference is reflected in practice. Of the fifty married women, including widows, among the Tsembaga in 1962 and 1963, twenty-two were Tsembaga by birth.

Some of the reasons for this preference have already been suggested. A man receives landrights from his wife's agnates. Only if this land is close by, however, may it be effectively used. Conversely, Tsembaga say that they prefer, other things being equal, to give sisters and daughters to local men so that even after her marriage a girl may continue to participate, to some extent, with her unmarried brothers or widowed father in gardening activities. Women and widowers, moreover, want at least one of their daughters to be married close by so that they may be assured of the funerary services that women perform.

Forty-one married women of Tsembaga birth were known to be alive during the fieldwork period. In addition to the twenty-two who were married to Tsembaga men, seven were married to the Tuguma. The remaining twelve had gone either to Jimi Valley men or to men north of the Simbai River.

Full satisfaction of a preference for local women is unlikely among groups the size of the Tsembaga. Even if the Tuguma, who number 225 people, are added, the population remains too small for easy local endogamy. Groups of such size are subject to imbalance in the numbers of persons in each sex available for marriage, especially when the sexes are subject to different mortality and, perhaps, different birth rates.[2]

While unions between men and women of a single local origin confer certain advantages upon both parties and their natal agnatic groups, unions between people of different local origins confer others. Unions across the grain of the land serve to strengthen trading relationships, for one thing. Perhaps more important to the welfare of the group is that allies are recruited through affinal ties. Thus, while it may be that the Tsembaga depend upon other groups for women for simple demographic reasons, the resulting interlocal ties enhance their ability to obtain commodities in trade and support in warfare.

Most marriages are arranged between the natal groups of the principals. In some instances an immediate direct exchange takes place: the two women are exchanged the same day. More common are delayed direct exchanges, in which a man, when he receives a woman, promises to reciprocate by sending to his wife's natal group at some future time a specified sister who is as yet too young to marry. A few women are given in what may be regarded as a trans-generationally delayed direct exchange. There is a rule stipulating that one of a woman's granddaughters is to be returned to her natal group. This is, in effect, a prescribed second patrilateral cross-cousin marriage (between a man and woman who call each other "brother" and "sister"). Such marriages are actually contracted in very few

2. The majority of nonlocal marriages are with members of other Maring-speaking groups, but four of the living wives and widows resident among the Tsembaga in 1963 came from nearby Karam-speaking groups. Conversely, two living women of Tsembaga birth were married to Karam men and another to a Gainj man. While no Tsembaga were married to Narak speakers prior to November 1963, marriage between Jimi Valley Maring and Narak speakers is frequent. It may be that propinquity, rather than linguistic affiliation, is the decisive factor in intergroup marriage. This question is likely to be resolved with the analysis of marriage data collected by A. P. Vayda during a pan-Maring census conducted in 1963 and 1966.

cases. Indirect exchanges are most common. The bride's natal group uses the wealth received from her husband to make payments on a woman procured elsewhere. Women are also sometimes given as compensation for certain services, such as revenging homicides, and figure in wergild settlements. This will be discussed later in the contexts of ritual and warfare.

Prestations must always be made by the groom and his kin to the bride's group even when the bride is given as wergild or in compensation for services. In most instances, however, no prestation, or only a small one, is made at the time the bride goes to the groom. It is said that the first substantial payment should be made soon after the bride and groom have harvested the first garden they have planted together; but it is frequently delayed much longer, sometimes until the next pig festival staged by the groom's group. Further payments are also made for both a woman and the children she bears throughout their lives and upon their deaths. The first marriage prestation is the largest, however, and may consist of over thirty wealth objects, as well as a whole cooked pig. Live pigs are never included in affinal prestations.

Wealth objects, or "valuables," form a named class, *muŋgoi,* which has traditionally included *meŋr* (shells: gold-lip, green sea snail, and bands of small cowries), *kabaŋ an* (feathers: most importantly bird of paradise plumes and parrot and eagle feathers), *ma wak* and *ma an* (animal pelts, used mainly as head bands, and loose animal fur used to embellish loin cloths), *čenaŋ* (cutting implements: steel axes and bushknives; previous to the arrival of steel, some stone axes[3] served only as *muŋgoi*) and *miña* (earth pigments). Austral-

3. Working axes are characterized by heavy, usually short blades, and are used for clearing, gardening, and fighting. Bridal ax blades, on the other hand, are long and sometimes extremely thin. One in my possession is 10″ long but nowhere more than .3″ thick. Such thin blades were not suitable for gardening, woodworking, or fighting, but were sometimes wielded during dancing, and they often figured in bride prices and other payments to affines. The distinction between the two types is recognized terminologically: bridal axes are referred to as *ambra poka čenaŋ* (woman payment ax), while a variety of working types are designated by their primary function (e.g. *ap čenaŋ,* tree felling ax). In fact, bridal axes are not separated from working axes by any sharp discontinuity. They form one end of a continuum with heavy working blades on the other. Intermediate forms both figured in bride wealth and were used for work.

While a variety of stones were used, *daŋunt,* black basaltic blades, were most common. *Gema,* blades of light green, grey, or white stone were also common (Chappell 1966).

ian money (*ku:* stone, including *ku meŋ*, stone pieces, and *ku wunt*, stone leaves), cloth (*a͂lap*) and trade beads (*budz*) have recently been incorporated into the *muŋgoi* class.

In the Simbai Valley in 1963 the items most prominent in affinal prestations were gold-lip and green sea snail shells and steel axes. Cowrie shell bands were going out of style but still were included, as were beads, bushknives, a few animal pelts, and occasionally pigments. Small amounts of money were included in two of the nineteen affinal payments that took place at the termination of the pig festival in 1963. Bird plumes figure in affinal prestations in the Jimi Valley, particularly among the neighboring Narak people (E. A. Cook, personal communication), but not in the Simbai Valley.

Although most marriages are the products of agreements reached by male members of the natal groups of the bride and groom, a girl's wishes in the matter are given serious consideration. Furthermore, it must not be imagined that the bride-to-be necessarily remains passive while her future is being arranged. It is often reported that girls refuse to go to the men whom their brothers or fathers select for them or, having gone, run away. Moreover, it not infrequently happens that girls run off to men without the prior consent of their male agnates. Fourteen, or 28 percent, of the fifty living married women and widows included in the sample were obtained by their husbands without the prior consent of their male relatives. This figure, which represents only those elopements that became enduring marriages, does not, of course, reflect the frequency of such occurrences, which must be much higher, for marriages are brittle previous to the birth of children. One Tuguma man who had two wives in 1963 reported that he had previously taken seven other women, only one of whom had died. Some of the remaining six were sent back to their natal groups by him and others left of their own accord. This number of unions for one man may approach the extreme, but other men also report several short-lived marriages.

Girls not only have considerable prerogatives in the final decisions concerning their marriages, but they also exercise considerable choice in the earlier stages of the process of mate selection. It is the female who initiates courtship, and young men attempt to make themselves appealing so that their advances will be invited. In this the *kaiko* plays an important part, as will be discussed in the next chapter.

Trade

A variety of goods comes to the Tsembaga through exchanges with other local groups and, in recent years, through exchanges with Europeans as well. While a number of other items figure in intergroup exchanges, most important during the period of fieldwork were bird of paradise plumes, axes, bushknives, European salt, shells, furs, and pigs. Previous to the arrival of Europeans and European goods, native salt manufactured from the waters of salt springs by Simbai Valley people, including the Tsembaga, and stone axes manufactured in two locations in the Jimi Valley were of great importance.

The importance of pigs in intergroup trade during 1962–63 was exaggerated by my presence. During that period the Tsembaga obtained in trade from other local groups 31 baby pigs (pigs above the age of three or four months are almost never traded). Most of these were obtained, however, with salt, face paint, or beads obtained from me in trade for food, or with shillings men earned by carrying supplies or equipment for me. It is doubtful if half that number would have been obtained in my absence. Of the 169 pigs constituting the herd at the beginning of the pig festival in June 1962, 56 were born elsewhere. Only 22 of these, or 13 percent of the herd, were obtained in trade from other places, however. The remaining 34 were brought to Tsembaga by their masters upon return from their exile following the military defeat of late 1953 or early 1954. To put it in a slightly different way, 87 percent of the pigs constituting the Tsembaga herd at the beginning of the pig festival were born on the territory of their masters.

Plumes and some furs come to the Tsembaga from northerly directions just as they did before their contact with Europeans. The Tsembaga themselves still add a few furs and plumes to the flow of goods, but in pre-contact times the most important contribution they and their Simbai Valley neighbors made to the trade was salt of their own manufacture. All of these items were sent to the south for commodities that came from that direction: pigs, shells, and both working and "bridal" stone axes.

Two of the items, native salt and stone working axes, are no longer produced by the Maring, but when they did figure in the exchange apparatus they were necessary or important to survival. Most of the other items were not. Bird of paradise plumes, fur headbands, shell ornaments, and bridal axes play no direct part in human subsistence. Neither do the minor items figuring in the exchanges: green bettles, pigments, and loose animal fur, all used to ornament the person, shields, or clothing.

Aesthetic considerations perhaps should not be overlooked in a discussion of this trade. To the Tsembaga and their neighbors, all of whom are without sculpture, ornamentation of the self or of shields is a form of artistic expression. They consider fine plumes and shells, gold-lip or green sea snail, to be among the most beautiful of objects, and men enjoy possessing them for their own sakes. They also, however, want the plumes so that they will be enticing to women when they dance, and they want the shells so that they can pay for the women they obtain.

It is also important that these valuables could be freely exchanged for stone axes and native salt. It may be suggested that the inclusion of both the nonutilitarian valuables and utilitarian goods within a single "sphere of conveyance" (Bohannan, 1955) stimulated the production and facilitated the distribution of the utilitarian goods.

Trade among the Maring is effected through direct exchanges between individuals. It may be questioned whether a direct exchange apparatus that moves only two or three items critical to subsistence would be viable. If native salt and working axes were the only items moving along a trade route, or were the only items freely exchangeable for each other, sufficient supplies of both might be jeopardized merely by inequities in production. Insufficiencies would develop because the production of each of the two commodities would not be determined by the demand for that commodity, but by the demand for the commodity for which it was exchanged. That is, the production of native salt would not be limited by the demand for salt, but by the demand for axes. If all that the Simbai people could obtain for their salt were working axes they would be likely to suspend the manufacture of salt if they had a large supply of axes on hand, regardless of the state of the salt supply in the Jimi Valley. The converse might be the case if the ax manufacturers had large stockpiles of salt.

It may be asked if the relationships of the people who regularly traded with each other would not in themselves be sufficiently important to the participants to compel them to supply each other's needs as a matter entirely separate from the fulfillment of their own needs (see Sahlins, 1965). In the case of the Maring salt–stone ax trade the answer is probably "no." The salt producers and stone ax producers were in almost all cases separated by at least two intervening peoples. A man may be able to bring moral pressure successfully to bear upon a kin or nonkin trading partner to supply his needs, or conversely, a man might feel morally bound to fulfill the needs of a trading partner. If, however, a man must put pressure on a trading partner to put pressure on a second, who will in turn put it on a third, who will attempt to get an ax from the manufacturers, success is less likely. It may be suggested that formal trading partner arrangements are sufficient to effect necessary exchanges in situations in which the trading partners exchange goods that they themselves have produced. But when a number of trade links separate the producers, other mechanisms may have to be employed. In other words, trading partnerships may adequately distribute needed exotic materials when each trader or each local group is the center of a *web-like* trading network. When, however, each trader or local group is a link in a *chain-like* exchange structure, the trading partnership by itself may not be sufficient to maintain an adequate supply of exotic materials. In such a situation the producers and consumers of a particular commodity may be separated from each other by a number of links and are usually unknown to each other. Moral pressure, which might induce production on the part of a trading partner, becomes too diluted between producers and consumers remote from each other to be dependable.

Although salt and working axes are both necessary to survival, the demands for them are limited. A man can eat just so much salt or use just so many axes, even if he is supplying some to kinsmen. If only salt and stone axes could be exchanged for each other the maintenance of a supply of both throughout the population would require a balance between the quantities of each produced and between their respective exchange values, both of which constantly fluctuate in response to such processes as local demographic changes. Without some kind of coordinating or directing managerial agency it seems hardly likely that such balance could be maintained. Where "redistributive systems" operate, that is, in populations in which supralocal authorities may demand production and enforce deliveries, exchange sets of such narrow scope may work. They might even work in "reciprocal systems" in which the parties to the transactions are *groups* in which production may be commanded by a local authority who might, conceivably, take into consideration the requirements of other groups. Among the Maring, however, the exchange apparatus was, and continues to be, based upon reciprocity between *individuals*, and no authority, local or supralocal, exists which may demand production or enforce deliveries.

The exchangeability of plumes, shells, and bridal axes with native-produced salt and working axes alters the relationship between these commodities. While the demands for salt and working axes were limited by the amounts required for specific processes of production, extraction, or metabolism, this is not the case with valuables. Plumes are perishable and there was, therefore, a constant demand for new ones. The size of bride prices is unspecified in advance, but a man is usually under pressure from his affines to pay well for his wife and is shamed if he is not able to do so. The unlimited demand for the valuables required to entice or pay for women, most importantly shells and "bridal axes," provided a mechanism for articulating the production of each of the two items critical to metabolism or subsistence to its own demand. As long as it was possible to exchange salt for shells or bridal axes, commodities for which the need was large but indefinite, salt would be produced. The production of salt would be suspended only with the suspension of the demand for it, and this demand presumably would reflect its status as a physiological necessity or near necessity. Conversely, as long as it was possible to obtain feathers for working axes, the latter would be produced. The demand for working axes, it may be assumed, would depend on the numbers of individuals gardening, the total area under production, and the kind of vegetation in which gardens were being cut.

Furthermore, the indefinite nature of the demand for valuables could serve to cushion fluctuating inequalities in the demand for salt and working axes. If, for instance, local population increases in the Simbai Valley increased demands for working axes at a greater rate than increases in the demand for salt developed in the Jimi Valley, the differences could be absorbed by an increased flow of valuables from the Simbai to Jimi Valleys. Any accumulation of valuables that the ax manufacturers might begin to amass would represent no threat to the continuation of ax production. Such accumulations might be drawn off by enlarged bride payments. They also might permit the ax manufacturers to obtain more wives. This could mean that women would move from the Simbai Valley to the Jimi Valley, which would suppress the birthrate in the Simbai Valley and augment it in the Jimi. Not only might the valuables have facilitated the distribution of axes and salt; they also might have thus provided a mechanism for distributing people more or less evenly over a broad area by adjusting differences in local population dynamics. Pig festivals also form an important part of this equalizing mechanism, as will be discussed in the next chapter.

No direct quantitative data in support of this interpretation of Maring trade are available, unfortunately, since the production of both native salt and stone axes ended some years before this fieldwork began. These suggestions may, however, serve as hypotheses that could be tested in areas where there has been less contact with Europeans.

HOSTILE RELATIONS

Hostile relations between Maring groups are characterized by long periods of ritually sanctioned mutual avoidance interrupted by armed confrontation or conflict. The last engagements in which Tsembaga participated as a principal combatant occurred in late 1953 or early 1954. Their enemy was Kundagai-Aikupa, the local population occupying the territory immediately to their west. Tsembaga men have taken part in warfare more recently, but only as allies or members or other local populations; on the most recent occasion, in 1958, they assisted the Tuguma against the Kanump-Kaur. This was among the last armed encounters to take place in the Maring area. Fieldwork did not begin until four years later, and information concerning fighting and the rituals associated with it therefore rests mainly upon informants' statements.

The Causes of Warfare

The proximate causes of Maring warfare may be inferred directly from accounts of fights and the events preceding them. The ecological-demographic conditions that may have underlain much of Maring fighting must remain hypothetical, since reliable quantified data concerning such conditions at the time of, and preceding, outbreaks of warfare are unavailable.

PROXIMATE CAUSES

Accounts of hostilities suggest that a distinction exists between events that originate hostilities between two groups and acts that merely maintain previously established enmities.

In every instance for which I have collected information, hostilities between previously friendly groups followed violence between members of the two groups. The violence, in turn, was induced in the particular instances reported by informants by (1) taking a woman without the prior permission of her agnates, (2) rape, (3) shooting a pig that had invaded a garden, (4) stealing crops, (5) poaching game and stealing scarce wild resources, and (6) sorcery accusations. Surely, however, there must be other acts that also have induced violence leading to intergroup fighting.

A violent episode between single or a few members of a pair of groups did not in all cases lead to intergroup armed conflict. Whether or not it did seems to have been mainly dependent upon (1) the results of the violence, and (2) the previous relationship of the two groups.

If the violence resulted in homicide, armed conflict between the groups of which the participants were members was more likely than if it resulted in mere wounding. Sometimes, however, homicides were settled peacefully and, conversely, less-than-fatal injuries occasionally led to intergroup fighting. The previous relationship of the groups to which the antagonists belonged is significant here. If the antagonists were members of different clans within a single local population it was likely that the trouble would be resolved without battle. When, for instance, the Tomegai and Merkai clans of the Tsembaga had a confrontation many years ago, members of the other Tsembaga clans stood between the shields of the antagonists, admonished them that it was wrong for brothers to fight, and implored them to break off the struggle, which they did.

That such fights, which are referred to as "inside fights" (*ura amaŋ*) or "brother fights" (*gui bamp*), were usually contained may well have been due to the great number of affinal and cognatic ties binding the several agnatic groups that form a local population. These ties not only provided a set of relationships through which composition might be attempted but also provided strong incentives for local neutrals as well as antagonists to seek settlement. It is likely that every, or almost every, local neutral, because of the high degree of local endogamy, would have close relatives among both sets of antagonists. If "brother fights" were allowed to proceed, these relationships would be seriously damaged. If armed hostilities were not speedily resolved, a pair of full brothers from a netural clan who had wives from the opposing antagonists would be likely to find themselves staring at each other across raised shields on a fighting ground, for military assistance is recruited through affinal connections. Even if the situation did not quite come to this, normal intercourse of those related to the antagonists might well be inhibited. There are taboos against eating food cooked over the same fire or grown in the same garden as that of one's enemy, and neutrals must make a choice. If the natal groups of the wives of brothers become enemies, either the brothers must choose between eating with each other or with their respective wives, or one or both of the wives must refuse to adopt the interdining taboos of her natal group. If a woman refuses to adopt the taboos of her natal group she may no longer dine with them. In any case, dyadic relations, which are heavily loaded both economically and with sentiment, are subjected to rather serious symbolic and behavioral impediments by unresolved disputes.

There are other incentives for the peaceful settlement of interclan disputes within the local population. It was shown in an earlier chapter that the local population as a unit exploits the territory. All individuals have access to nondomesticated resources existing anywhere on the territory, and garden land is intermingled. A dispute between the subclans of a clan, or between two clans within a single local population could, if it were not settled or at least contained, make it necessary for the population to redisperse itself radically over these resources. Members of the groups involved know that parcels of land upon which they are presently gardening or upon which they plan to garden would have to be abandoned in case of a continuing dispute; usually they do not desire such a redispersal and are anxious to see any hostility resolved.

Furthermore, the local population forms a single military formation, which faces the enemy as a unit. Intrapopulational disputes debilitate such a unit and may even diminish it in size, for it sometimes happens that one of the parties to a dispute departs from the local population. Such a loss may jeopardize the position of those remaining in later confrontations with the enemy. As the maximum size of local populations is limited by the carrying capacities of their territories, so are minima set by the size of the groups whom they confront as enemies. This principle seems not to be lost upon the Maring: depleted groups have several times invited people from elsewhere to join them on their land. Informants reported that three such events occurred within the Tsembaga and Tuguma local populations during the fifty years previous to fieldwork, and Vayda and Cook (1964:801) suggest that such invitations were common throughout the Maring and Narak areas.

Local populations may also be regarded as congregations, in that the performances of certain important rituals by constituent subterritorial groups are coordinated, and it may be that the Maring regard the necessity to continue performing these rituals together as another imperative for resolving disputes within the local population.

Peaceful settlement of disputes is less likely between friendly adjacent groups that do not form a single local population, that is, do not share rights in the nondomesticated resources of a combined territory, do not form a single military unit in confrontations with the same enemy, and within which all constituent units do not plant and uproot their *rumbim* at the same time. While there is considerable intermarriage between such adjacent groups, it is usually less than that which takes place within each of the local groups; there are thus fewer channels through which peace may be effected and less incentive to do so than is the case within the single local population. Moreover, disputes between adjacent but separate local populations endanger neither patterns of dispersion over resources nor military capability to the same extent as intrapopulational strife. While there may be some intermingling of garden land between adjacent local populations, unless they are in the process of fusion it seldom reaches the proportions that it does within single local populations. And while the adjacent groups act as allies to each other in the military encounters that each have with their separate enemies, there is a difference between the ways in which principal combatants and allies participate in fights.

It may be said, thus, that fighting between separate territorial groups is more likely to occur than fighting between descent groups that are constituents of a single local population. Settlement of disputes within the local population is more likely because of the mutual interests not only of the antagonists but also of other agnatic groups forming part of the same local population. Furthermore, all of the members of the local population are closely bound to each other in a web of kinship connections that serve as channels through which problems may be solved. "Inside" or "brother" fights do occur and sometimes lead to fission and the adoption of formal enmity relations, but we shall be concerned here with fighting between separate local populations, not internal fights.

Once an enmity is established, the requirements of vengeance play an important role in maintaining it. A principle of absolute reciprocity is supposedly in force; every death at the hands of an enemy group requires the killing of one of that group's members, and peace should not be made until both antagonists have revenged all of their losses. A series of military engagements may be terminated, even though the homicide score remains unbalanced, but only by a truce, with arms to be taken up again in the future. The periods of truce may last for ten years or more, however, depending upon a number of factors, which will be discussed in the next chapter.

An even score in killing is not easy to achieve, even though for purposes of vengeance all men, women, children, and babes in arms are considered to be equivalent, and most military episodes terminate with blood debts remaining. Since this is the case, each round of fighting contains within it the seeds of the next. The Tsembaga–Kundagai-Aikupa hostilities in 1953, for instance, were set off by an attempt on the part of some Tsembaga to erase a blood debt by killing a Kundagai. This blood debt was the product of the previous series of engagements, perhaps ten or twelve years earlier, which in turn had been set off by blood debts arising out of a yet earlier conflict.

Some practices may act to preserve the principle of homicidal reciprocity while mitigating the rigorousness of its application. A distinction between the responsibilities of allies and those of the principal combatants has such an effect, as may certain magical and religious procedures, which will be discussed later. Here we are concerned with the ways in which the Maring assign responsibility for revenge and credit for homicide and how they may serve to temper the requirement of absolute reciprocity. A clan is responsible for the vengeance of deaths sustained by its own members and their wives. If in the course of a battle an enemy is brought down, it is usually the case that a number of men participate in administering the coup de grace. Each, claiming credit for the homicide, may regard vengeance accomplished for a member of his own clan. One corpse, thus, may fulfill the reciprocal requirements of several homicides.

On the other hand, the way in which responsibility for the initial homicide is assigned serves to extend and exacerbate enmities and, perhaps, to increase killing. The agnates of the victim of homicide hold as responsible not only the agnatic group of the slayer but the entire local population of which the slayer is a member. Vengeance can be wreaked against any member of this larger group. A previously uninvolved agnatic group may thus become embroiled in a vendetta by suffering revenge for a killing that was not committed by any of its members. Obliged to repay such a homicide, its members will seek vengeance against the local population of the slayer, perhaps victimizing a group that was not party to the aggression they themselves had suffered.

UNDERLYING CAUSES: POPULATION PRESSURE

Although data are deficient, and informants deny fighting over land, there are indications that at least one of the fights in which Tsembaga have been involved was a response to population pressure.

No census material is available for the period during and immediately preceding a fight between the Tsembaga, primarily the Merkai clan, and the Dimbagai-Yimyagai thirty to forty years ago. It was estimated in Chapter 2, however, that the Tsembaga population probably stood between 250 and 300 persons at the time. Estimates also suggest that such a number would have been pressing upon the carrying capacity of a territory, which was at the time smaller than in 1962–63. People say, furthermore, that at the time of the fight there were many people and not much land.

The circumstances surrounding the fight itself also suggest that land shortage was the cause. The dispute that may be regarded as the proximate cause of the fight should have been easily resolved since it did not result in homicide. The garden of a man named Paŋwai, who lived with the Dimbagai-Yimyagai, lay close to the path that Merkai men followed to their trapping and collecting grounds in the high forest. Located at about five thousand feet, it was the last garden one would pass before entering the forest and the first one would come upon on the way home, and men often rested by it and slaked their thirst with its sugar cane. Paŋwai was annoyed by what he regarded to be their thievery, and when he caught Kati "drinking" his sugar cane, he shot him. Kati was not seriously injured: the wound was in the buttock and the arrow was unbarbed. Nevertheless, the Merkai took up arms against the Dimbagai-Yimyagai. It was their position that Paŋwai would have been correct to shoot a man whom he caught taking taro from his garden, for to take taro is indeed stealing. But Kati, who had taken only a little sugar to quench his thirst, did not deserve to be shot.

A striking aspect of the affair was the agnatic affiliations of the principals. While Paŋwai was living with the Dimbagai-Yimyagai, he was actually a Merkai by birth. Although Kati was living with the Merkai, he was actually a Karam-speaking man living with the husband of a sister married into Merkai. No Dimbagai-Yimyagai was party to the dispute. The Merkai could have regarded the incident as an intraclan dispute rather than an interlocal population matter. Moreover, they were not required to avenge Kati, for Kati was not an agnate.

Another point of interest is that at the time the fight broke out there were nine Merkai men living and gardening uxorilocally, sororilocally, or matrilocally upon the territory of the Dimbagai-Yimyagai, while only one Dimbagai-Yimyagai man was gardening on Merkai territory. Informants, while they do not make clear why these Merkai did not maintain patrilocal residence while gardening on Dimbagai-Yimyagai ground, state that their reason for gardening on ground other than that of their own clan was land shortage.

These men, it should be noted, fought on the side of the Dimbagai-Yimyagai against their own agnates. This indicates either that they had accreted to the Dimbagai-Yimyagai or, more likely, that the Dimbagai-Yimyagai and the Merkai were fusing when fighting broke out between them. The possibility of fusion being aborted is consistent with the density-dependent sequence of land amalgamation and structural arrangements suggested earlier (page 27). At moderately high densities the lands of adjacent groups become intermingled through affinal and then cognatic grants, which serve to even out the dispersal of persons over wider areas than is possible through reliance upon unilineal descent principles alone, and the process of fusion proceeds through the stages of the affinal and then the cognatic cluster. However, if densities become very high during this process and an absolute shortage of land develops in the total area occupied by the fusing groups, it is reasonable to expect that the fusion process will cease, and fighting, leading to the expulsion of one of the previously fusing groups, is perhaps likely.

Even less information is available from which we would or could infer environmental or demographic conditions at the time of other fights. It may be suggested here, however, that the frequencies of some kinds of disputes are dependent on population densities.

If twenty men, for example, each own one pig and have one garden, there are 400 possibilities for pigs to cause disputes between men by damaging gardens. If the number of men is raised to forty, each of which still has one pig and one garden, the number of possibilities for disputes has increased to 1,600, other things being equal. Likewise, doubling the numbers of unmarried males and unmarried females also, perhaps, more than doubles the possibilities of woman stealing and other dispute-producing incidents. Sources of irritation thus increase at a greater rate than population size. If population increase were taken to be linear, the increase in some causes of dispute, if not actual dispute, might be taken to be roughly geometric. It might even be possible to find some way to express mathematically an "irritation coefficient" of population size.

This increasing irritability should begin to express itself in disputes well below the actual carrying capacity of an area. However, there is little information relevant to this hypothesis concerning the Tsembaga, even of an anecdotal sort. The earliest fight about which Tsembaga informants have any clear recollection took place fifty or sixty years before this fieldwork, when the Kamuŋgagai and Tsembaga clans together fought the Kekai, a group living across the Simbai River. Informants say that all of the parties to the fight had sufficient land.

Previous to this, in the time of their grandfathers, middle-aged and old informants say, fights were infrequent and enmities were brief, and there are also indications that the population was smaller. There is some evidence, although equivocal, in genealogies, in the flora, in clan histories, and in accounts of the removal of virgin forest from extensive tracts below 4,500 feet within the past sixty years, to indicate that the Tsembaga and their neighbors have arrived in the Simbai Valley within the past two hundred years, that they were relatively peaceful for a long while after their arrival, and that when intergroup fighting began, the population had already expanded considerably but had not yet approached carrying capacity.

It is possible that increasing irritability accompanying a growth in population size might serve, if some of the disputes result in the movement of individuals or groups, to protect the environment from degradation by continually redispersing populations over available land before the carrying capacity is reached.

Composition of Fighting Forces

Fighting forces almost always include two categories of men. The first may be referred to as principal combatants. These are members of the local populations that are parties to the dispute from which the armed conflict has arisen. The second are allies—members of other local populations who are recruited by the principal combatants through cognatic or affinal ties. In some instances men also give armed assistance to nonrelated trading partners.

Because of frequent intermarriage between the Tuguma and Tsembaga through a number of generations, almost every member of each has affines or cognates in the other. The result is, in effect, that the Tuguma as a unit assist the Tsembaga in their fights against the Kundagai, and the Tsembaga as a unit come to the aid of the Tuguma in their encounters with the Kanump-Kaur. While it may be the case that Tuguma-Tsembaga form a single military unit for the defense of a continuous area, the rationale of recruitment of one by the other remains that of kinship ties between individuals.

Kinship ties are less numerous between groups separated by mountain ridge or river. The result is that military assistance rendered by these geographically more distant groups to each other is on a smaller scale, consisting of men who are closely related to the principal combatants and, if these men are influential, some of their relatives as well.

Principal combatants are responsible for casualties sustained by their allies,[4] and they must compensate the agnates of a slain ally with, among other things, a woman, whose first son will bear his name. The practice of assigning the responsibility for casualties sustained by allies to the principal combatants whom they assist, rather than to enemies, may serve to diminish the frequency of fighting throughout the Maring area. It frequently occurs that enemies of long standing find themselves confronting each other as allies of other principal combatants. The Kanump-Kaur, for instance, who are the enemies of the Tuguma, have many affinal connections with the Kundagai and therefore turn out in large numbers to assist the Kundagai in their fights against the Tsembaga, who are assisted by the Tuguma. The Tuguma and Kanamp-Kaur face each other in such situations not as principal combatants but as allies of opposing principal combatants. If in the course of such an encounter the Tuguma slay a Kanump-Kaur, they may regard this as revenge for a previously slain Tuguma. The Kanump-Kaur, on the other hand, would not charge this slaying to the Tuguma, but look to the Kundagai for redress. If such a killing fulfills the revenge requirements of a group it will not be necessary for it to seek vengeance on its own and set off another round of warfare.

Fighting and Its Rituals

Accounts of fighting and the rituals associated with it sometimes conflict. This is true particularly with respect to the sequence and details of prefight ritual procedures. Differences in accounts may in part be explained by variations in the practice of different groups and even different men. A simplified account of ritual performances is sufficient here; variations will be discussed elsewhere. What follows reflects the practice of two of the three Tsembaga subterritorial

4. Glasse (1959) reports very similar assignments of responsibility among the Huli people of the Southern Highlands of New Guinea.

groups; the practices of the third differ in some details, according to informants.

THE "SMALL" OR "NOTHING" FIGHT

Most fights have two stages, which are distinguished by the weapons employed and the rituals performed.

The first of the two stages is referred to as the "nothing fight" (*ura auere*), or "small fight" (*bamp ačimp*). A local population that has suffered injury calls out to the enemy to prepare for an encounter on a designated fight ground. One or two days' notice is usually given, providing ample time for both sides to mobilize their allies and to clear the fight ground of underbrush. This task falls equally upon both antagonists but encounters between the bush-clearing details of the hostile groups are avoided. Informants say that if members of one of the groups arrive at the fight ground to find their opponents already at work they will withdraw for some distance until their opponents retire, after which they will enter the fight ground and finish the clearing.

Prefight rituals are performed separately by the three separate Tsembaga subterritorial groups. On the night previous to a "small" or "nothing" fight, informants say, the warriors convene by clans or subclans in men's houses to inform the spirits of both the high and low ground of the next morning's encounter. As they sit chanting in a darkness illuminated only by the embers of small fires, shamans (*kun kaze yu*), of whom there are several in each clan, induce in themselves an ecstasy by inhaling deeply and rapidly the smoke of bespelled cigars made from strong native tobacco.[5] When his *nomane* (his animated, immortal, thought stuff) departs through his nose to seek out the smoke woman (*kun kaze ambra*) in high places, the shaman begins to tremble and gibber. Soon the smoke woman "strikes" him. Led by his *nomane* she enters the shaman's head by way of his nose, and his ecstasy reaches its height. Rising to his feet he dances about the embers in a low crouch, sobbing, chanting, and screaming in tongues. It is through the smoke woman that the ancestors are now being informed by the living of the fight and it is

5. While I have not witnessed the performance of shamans in connection with fighting, I have seen shamans perform in a number of other situations. Informants assure me that their performances in the context of fighting are similar to those I have witnessed.

through her that they are now signifying their endorsement of the enterprise and sending assurances of their protection. It is, informants say, the smoke woman who speaks in tongues through the shaman's mouth.

The protection of the ancestors is bestowed upon the warriors through the shaman's body. As he dances within the circle of men, the shaman takes each of their extended hands in turn, and with it wipes sweat from his own armpits. This sweat, driven from his body by the heat of the entering spirit, is said to be hot and therefore imparts strength. No pigs are killed at this time, informants agree, but slaughter is promised in the event of a satisfactory outcome of the fight.

It is not clear whether the smoke woman is asked at this time to name those members of the enemy group who may be easily killed and to warn those members of the local group who are in special danger. Some say that this forms part of the ritual preceding the more deadly second stage of fighting only.

In the morning spells are said by "fight magic men" (*bamp kunda yu*),[6] of whom there are one to three in each clan, and who are usually, but not necessarily, shamans as well. The spells are said over the bows and arrows so that they will be strong, accurate, and sharp, and over the shields so that they will stop the arrows of the enemy. Small bundles, called "fight packages" (*bamp yuk*) or "fight bags" (*bamp kun*), are pressed to the hearts and heads of each warrior by the fight magic man. These packages contain the thorny leaves of the males of a rare, unidentified tree growing in the *kamuŋga*, called the "fight tree" (*ap bamp*), and personal material belonging to the enemy, such as hair, fragments of leaves worn over the buttocks, and dirt scraped from the skin. It is said that pressing the "fight leaves" (*bamp wunt*) contained in the packages to the heart and head of a man diminishes his fear, and that control over the personal items of enemies, and their confinement within the same package as the leaves, enhances the opportunities for killing those who are represented. These materials are obtained from neutrals who have kinship connections among the enemy and therefore may visit

6. I have used *magic* as a translation for the Maring term *kunda*. *Kunda* refers to procedures involving the repetition of conventional formulae, or spells, that are said to derive their power from the words themselves, rather than from any other agency. Spirits are not invoked in any of the *kunda* spells I have collected.

them. A neutral himself may, if he suspects a man of being a witch (*koimp*) or sorcerer (*kum yu*), acquire such objects by stealth, but it is said that sometimes a man's own clan brothers, suspecting him of witchcraft or sorcery, will give some of his exuviae to a neutral to deliver to the enemy.

While one fight magic man is pressing the fight package to the hearts of all men, another is ritually applying a gray clay (*gir*), obtained from the *wora*, or low ground, to their legs so that they will remain strong throughout the day.

When treatment of both the heart and legs is completed, lengths of green bamboo are laid on a fire. With the warriors gathered in a circle around them, one or more fight magic men kneel by the fire and talk to the ancestors, sobbing as they implore protection. When the bamboo explodes, the men stamp their feet and, crying "ooooooo," leave in single file for the fight ground, carrying shields, bows, and arrows. They sing a song called "*wobar*" as they proceed in a peculiar prancing gait toward battle.

In the small fight only bows and arrows, and perhaps throwing spears, are used. Some informants say that hand-to-hand weapons such as axes and jabbing spears are not even brought to the fight ground.

The antagonistic groups line up on the fight ground within easy bow shot of each other, according to informants, who also say that allies and principal combatants are intermingled in the formations. The shields, which are very large, averaging 2.5 by 5 feet, are propped up, permitting bowmen to dart out from behind them to take shots and leap back again. To demonstrate their bravery, men also emerge from behind the shields to draw enemy fire. Casualties are not numerous and deaths infrequent, for the unfletched arrows of the Maring seldom kill.

It may be that the "small" or "nothing" fight is less a serious battle than a device for ending a quarrel before more lethal fighting, with its attendant ritual constraints, develops. The relative harmlessness of a bow and arrow engagement from stationary positions with cover well provided by shields itself suggests such a possibility, but there are even more cogent reasons for seeing the small fight in such a light.

First, small fights are protracted. Accounts of engagements in the past indicate that in some instances there were four or five days of such fighting before escalation took place. Such a period might allow tempers to cool while satisfying the bellicose imperatives of manhood.

Second, while there are no third parties with either the power or authority to adjudicate disputes, the formations of the antagonists include allies, men less committed to the quarrel than the principal combatants. These men often have a considerable interest in seeing the quarrel settled before it goes any further, for they themselves have no direct grievance and may have close relatives in the enemy formation. When the Tuguma fought the Tsevent, for example, some Tsembaga went to the assistance of both of the antagonists. In one instance, this split two full brothers who had sisters married into each of the opposing groups. Although they take their places in the military formations of the antagonists, such men may serve to dampen the martial ardor of the principal combatants by obliquely counseling moderation. Allies seem to have behaved in this way during at least one confrontation between antagonists who had previously been friendly. During the Tuguma-Tsevent fight, informants say, some of the Tsembaga allies of both antagonists, instead of hurling insults at their opponents, lamented loudly and continually about the evil of brothers fighting. Their laments in this instance were unheeded, perhaps because two or three Tuguma are reported to have been killed by arrows during the little fight. If there had been no fatalities, or if only one Tuguma had been killed in reciprocation for the homicide that had led to the fight, the preachings of the Tsembaga may have had more effect.

Third, the small fight affords an opportunity for nonallied neutrals to attempt to intervene between antagonists who have previously been friendly. Reference has already been made to the fight between the Dimbagai-Yimyagai and the Merkai clan of the Tsembaga. At the initial engagement neutrals stood on a knoll, informants say, overlooking the fight ground, admonishing both sides that it is wrong for brothers to fight, demanding that the combatants quit the field and throwing rocks at both formations. Their efforts were unavailing, it must be reported, and the fight escalated.

Finally, the "small fight" brings the antagonists within the range of each others' voices while keeping them out of the range of each others' deadlier weapons. Informants say that this opportunity for communication was used mainly to hurl insults at each other, but there is some evidence to indicate that in some fights, at least, the opportunity was used to resolve the quarrel. Information on this aspect of the small fight is ambiguous. There are several reported instances of small fights being terminated by shouted negotiations after one side had sustained a fatality that the other side could regard as full reciprocation for a homicide it had sustained and which had set off the fight. In such instances the termination of the conflict should be attributed to the fulfillment of revenge requirements, rather than to the negotiations, but the small fight at least provided an opportunity to fulfill the revenge requirements without recourse to deadlier fighting, which would have been more likely to compound grievances than to cancel them.

When information is derived mainly from the war stories of old campaigners it is difficult or impossible to know what actually occurred. There are, nevertheless, indications that the small fight operated, albeit in an inefficient way, to suppress rather than to encourage hostilities. Such an interpretation acquires additional plausibility when the small fight is compared to the deadlier encounter that usually follows it, and when the process of escalation is examined.

ESCALATION

Informants' accounts indicate that from the first day of the small fight, or certainly after one or two days, some members of one or both of the antagonistic groups would begin to shout to both their opponents and their comrades that they had had enough of the "nothing fight" and that the time had come for a serious engagement. It may be that such sentiments formed a kind of counterpoint to the laments of allies, in the case of fights between erstwhile friends, for the lost peace. The small fight may perhaps be viewed as a debate, held in a setting that minimized the danger of casualties while satisfying martial imperatives, between those eager to fight and those hoping to preserve the peace. All the interested parties, both the antagonists and their less-committed allies, were present and in a position to contribute opinions and sentiments.

Furthermore the small fight is a show of force. The antagonists are given an opportunity to assess the strength of their enemies and to shape their policy accordingly. There is little information on this point, but an apparent equivalence of forces may encourage settlement of the quarrel. Conversely, an apparent disparity may induce the weaker group to flee without putting the matter to a further test of arms. Some informants say that this is what happened in the second round of engagements between the Dimbagai-Yimyagai and the Merkai, but agreement among informants is not complete.

As a show of force, the small fight may resemble in certain respects the war games and military displays of more complex societies. It bears an even closer apparent resemblance, however, to the agonistic territorial displays of other animal species and may well have had an effect similar to that which some scholars (e.g., Wynne-Edwards, 1962) attribute to this kind of animal behavior: maintenance of optimum population dispersal at minimum expense in blood.

The success of such exercises in diplomacy in avoiding escalation depended largely on two factors. These were, first, the previous history of the relationship between the antagonists. If their hostility were long standing or if the homicide score between the two were seriously out of balance, those eager to fight would no doubt prevail. Also important was the extent to which the events of the bow and arrow fighting exacerbated or assuaged hostilities. It may be that if casualties were few tempers would cool and voices would be added to those speaking in the cause of peace.

Eventually the talk would, as the Tsembaga say, "become one." If those in favor of peace prevailed, the fight would be dropped, and if the antagonists had been friendly until this engagement, they could resume their friendship without going through elaborate ritual procedures.

However, it was frequently, perhaps even usually, the case that a consensus in favor of fighting crystallized. If it did, the antagonists withdrew from the fight ground for at least two days to prepare ritually for the forthcoming encounter.

THE AX FIGHT

The second, more serious stage of the fight has no native designation; it is simply called by Tsembaga the "*ura kuňuai,*" "fight itself," or "true fight." It will be referred to here as the "ax fight," although jabbing spears, throwing spears, and bows and arrows are also employed.

The rituals preparatory to the ax fight are much more elaborate than those preceding the small fight. In the last chapter it was mentioned that each land-holding group periodically plants a ritual *Cordyline fruticosa* bush called the "*yu miñ rumbim*" or "men's souls *rumbim.*" [7] A group may not participate as principal combatants in ax fights with its *yu miñ rumbim* in the ground. There are periods when land-holding groups do not have *yu miñ rumbim* growing, and fighting is generally confined to such periods (this will be discussed in the next chapter). It occasionally has happened, however, that a group who has uprooted its *rumbim* has initiated hostilities while the *yu miñ rumbim* of its opponents has remained in the ground. If a group has suffered an attack while its *yu miñ rumbim* continues to grow, it must, before proceeding to the other rituals preparatory to the ax fight, uproot this plant. The elaborate rituals that surround the uprooting of the *rumbim* under peaceful conditions are abbreviated when a group is forced into the procedure by enemy action. Description of the procedure will be reserved for the discussion of the *kaiko,* the context in which it more frequently occurs.

After the *rumbim* is uprooted, the ritual hanging of the "fighting stones" (*bamp ku*) is begun. Each of the land-holding groups owns at least one pair of "fighting stones." These are also sometimes referred to as "*aram ku*" (the meaning of *aram* is obscure). These stones are in the care of one or two of the fight magic men, who are referred to as "*aram* stone men" (*aram ku yu*) or, because of the large number of ritual restrictions under which they labor, "taboo men" (*aček yu*). Fighting stones, which are in fact stone mortars and pestles made by a vanished and unremembered people, are occasionally found in the ground in both the Jimi and Simbai Valleys. Informants believe, however, that all of them originated at the first home of the smoke woman, Mount Oipor, in the upper

7. In contrast to the *nomane,* which is immortal, the *miñ* is lost at death. Conversely, its abandonment of the body, an occurrence that may be induced by sorcery, bewitchment, or fright, results in death. The *miñ* is thought to walk abroad in dreams, and the moment of waking is particularly dangerous, for a sudden awakening may not allow it to return to the body. The *miñ* may, in short, be regarded as mortal life stuff.

Jimi Valley, and say that their ancestors obtained them from people in that vicinity.

The fighting stones are ordinarily kept in a small, round house called *riŋgi yiŋ* (*riŋgi*/fighting ash; *yiŋ*/house). During the portion of the ritual cycle preceding hostilities, the stones are kept in a small net bag on the floor of the house, where they also remain during the small fight. When the ax fight is agreed upon, the net bag is taken from its resting place and hung near the top of the center pole. Informants say that this is done by the taboo men unceremoniously, but it is a momentous act. By hanging up the fighting stones, a group places itself in a position of debt to both allies and ancestors for their assistance in the forthcoming ax fight. These debts can be fulfilled only through the prolonged ritual procedures that will be discussed in the next chapter.

The act of hanging the fighting stones signifies the assumption of a large number of taboos by the entire local population or various of its members. These include, for all members, females and children as well as adult males, prohibitions against trapping marsupials, eating eels, or eating marsupials and *marita pandanus* fruit together. Adult males may not eat the flesh of pigs killed for the *kaiko* of other groups, many kinds of marsupials, some yams, and a number of green vegetables. A number of other foods may be eaten by both men and women, but not shared by them. Drums may not be beaten, and the movements of some men are severely restricted. Some informants say that the taboo men may not depart from the territory of the local population, although others say they may visit the adjacent Tuguma, but not groups whose territories lie across the Bismarck Mountain ridge or the Simbai River. Furthermore, the fires over which their food is prepared, or which serve to heat stones for the ovens from which they partake, may not be used to cook food for members of other local populations.

Most important, the opposing group, which may have been until this moment referred to as "brothers" (*gui*), if they had previously been friends, now become, formally, "ax men" (*čenaŋ yu*), or enemies. Their territory may no longer be entered except in battle, and outside of the context of battle members of the enemy group may not be touched or even addressed. One is not even supposed to look at their faces. No cultivated food grown on enemy ground may be eaten, nor may food grown elsewhere by a member of the enemy group. Should fatalities be suffered, food grown by allies of the enemy must also be avoided, and, while they may be addressed and even visited, their houses may not be entered. If no fatalities are suffered, food that allies of the enemy have grown may be eaten, but dining from the same fire with them is forbidden. The act of hanging the fighting stones, in short, not only terminates any kind of mutually supportive relations between the principal combatants, but also places restrictions upon the future relations of principal combatants with the allies of their opponents. Moreover, some prohibitions, particularly in the area of food sharing, affect relations between allies of the opposing principal combatants if they suffer fatalities.

Hanging the fighting stones, furthermore, also pushes far into the future any possibility of reconciliation between the principal combatants. Truces are possible while the fighting stones hang, but reconciliations are not. The time for quick settlement of the dispute has passed.

After the stones have been hung on the center pole one of the fight magic men climbs to the roof of the *riŋgi* house, where he ignites a stick of wood taken from a tree called *"kawit"* (*Cryptocarya* sp.). This is obtained from the dwelling places of the red spirits near the top of the Bismarck Mountain ridge. *Kauit* is one of the varieties of trees said to be the houses of the red spirits. It is particularly difficult to ignite, and the fight magic man sobs as he recounts to the red spirits the circumstances that have led to this performance and implores them to come into the fire. When he succeeds in igniting the *kawit* he returns to the ground and uses it to ignite a fire that has been laid inside the *riŋgi* house to heat oven stones and in which the ritual ash, *riŋgi,* will be prepared from logs of *kanam* (*Albizzia* sp.), *kamukai* (*Colona scabra*), and *pokai* (*Alphitonia iacana*). These second-growth trees have themselves no ritual significance; they are used, informants say, only because their combustion produces good black charcoal. Their ash, or charcoal, is supernaturally powerful only because the red spirits themselves have been brought into the fire upon which it is prepared. The oven stones also partake of the supernatural power of this fire, and the fighting stones, always strong, are fortified by the smoke surrounding them, as they hang from the center pole.

A second fire is laid outside the house for the purpose of heating cooking stones for a second oven. This is ignited at the same time as the indoor fire, but not with the *kawit.* The red spirits do not enter or ignite this fire, and no *riŋgi* is prepared in it.

Two pigs are now slain. One of them is to be offered to the red spirits and must be a male. The second, to be given to the spirits of the low ground (*the rawa mai,* a category, it will be recalled, which includes both *rawa tukump,* spirits of rot, and *koipa maŋgiaŋ*), may be either male or female.

When pigs are sacrificed to them, spirits are usually addressed in a peculiar screaming style. The message is delivered in staccato phrases, interrupted with increasing frequency by meaningless, loud, sharp yells until, just before the pig is struck on the head with the club, the staccato "Ah! ah! ah! ah! ah!" has replaced words completely. Informants say that the addresses to the spirits in connection with the killing of the pigs in preparation for the ax fight follows this pattern, which I have observed in other contexts on several occasions.

The red spirits are first approached. They are not addressed as *rawa mugi.* Instead a series of allusive terms referring to their characteristics and activities, and pseudonyms that may also be allusive but remain obscure, are used. Allusive designations include "sunfire" (*ruŋga-yiñe*), referring to their hotness and, therefore, strength; "orchid, cassowary" (*norum-kombri*), referring to their habitation of high places and fighting qualities, and *niñ niñ koramon,* which designates the act of lighting the fire on the roof of the *riŋgi* house. After being addressed as a class, individual red spirits—the slain brothers, fathers, and perhaps grandfathers of the living—are addressed and told to bring their associates to watch and listen.

Having alerted the red spirits, the slaughterer, who is one of the fight magic men, proceeds with his address. The burden of his message is to tell the spirits that an ax fight is about to start. He promises that the living will attempt to avenge their deaths and asks their help. He invites them to eat the pig that is about to be killed for them and to come into the heads of the living. He promises that only members of the local population will join them in eating the flesh of this pig. Men from other places, he tells the red spirits, will be given other pigs.

His short address finished, the screaming slaughterer runs back and forth in front of the pig, which is tethered by one leg only, and strikes the animal with his club while on the run. The first blow usually kills the pig.

The spirits of the low ground are approached in a similar way, although they are addressed neither as *rawa mai* nor *rawa tukump*, but as ancestors (*ana-koka,* literally, father-grandfather) and as *koipa maŋgiaŋ*. These spirits are also told of the forthcoming encounter, and they are invited to eat the flesh of the pig offered them and to take hold of, or strengthen, the legs of the living.

The pigs are then placed in the ovens for cooking. The pig that has been dedicated to the red spirits, sometimes referred to as the "head pig," is cooked in an oven built entirely above the ground out of wood and leaves. Stones heated on the fire made inside the *riŋgi* house, which had been ignited by and entered into by the red spirits, are used in this oven.

After the pigs have been put into the ovens, the men retire into their houses. They spend a substantial portion of the night chanting while the shamans contact the smoke woman, whom they ask to name those enemies who may be easily killed and to warn those of the local men who are in especially great danger. It is usually the case, informants say, that only one or two enemy men are designated easy marks in each of the three to six Tsembaga men's houses in which shamans perform. Similarly, only a few of the locals are warned. Informants say that a warned man is urged by his fellows to stay away from the fight ground, but his usual response is to refuse, saying that he will kill one of the enemy first, after which it is all right if he himself is killed. He may add that it makes little difference if he is killed, for many clan brothers will survive him. If a warned man survives the fighting, his reputation is enhanced, for his *miñ,* his life essence, is said to be strong.

A second procedure is also used to mark those of the enemy who may be easily killed. A separate, small above-ground oven is built in the *riŋgi* house to cook the head of the pig killed for the red spirits. During the night, informants say, the *miñ* of some of the sleeping enemy, which wander abroad while they dream, may be lured by the pleasant odor of cooking pork into the *riŋgi* house, where they will partake of the pig's head. Such men may be killed. The taboo man sits by the oven and recites the names of enemy men. The head of the pig, it is said, whistles when the name of an enemy whose *miñ* has partaken of it is enunciated. The whistling of the pig's head is said to be audible not only to the taboo man but also to others, most of whom cannot be in the *riŋgi* house, but are rather in nearby men's houses or, less likely, out of doors.

Several aspects of these two types of divination ritual should be discussed. It is of interest that only a few of the enemy are marked for easy killing by the two procedures. It may be that this is simply the expression of a realistic appraisal of the situation, i.e. that it would not be easy to kill a larger number. There is some suggestion, however, that the number designated, although not large, represents "sufficient" homicides—enough, that is, as far as the local group is concerned, to halt the fighting. This is not necessarily a reflection of the "homicide score" existing between the antagonists, for the group that has suffered fewer fatalities in the past also goes through such divining procedures. The data are far from sufficient, but it may be that such divination sets up a rough "killing quota." The possibility that such quotas are established through ritual or other means could be explored by anthropologists working in areas where warfare still occurs.

The designation of certain enemies as easy marks also may serve to direct the homicidal attentions of the entire force to them. That is, such prophesies in effect may be self-fulfilling. If this is the case, both shamanistic practices and pig's head divination may be constantly reinforced by their apparent successes.

Clear information relevant to this matter can no longer be obtained. Conversations with shamans would indicate that many of the enemy who are marked actually are slain in battle. Shamans, however, boast about their own successes and, in confidence, disparage the abilities of other practitioners. The actual facts of particular cases cannot be known for certain.

It may be that the use of enemy exuviae, combined with the practice of predicting who among the enemy are easy prey, may act as part of a social control mechanism. As in the matter of the accuracy of predictions made by shamans, data are deficient concerning fight packages because their owners, while sometimes boastful about their efficacy, are always secretive about their specific contents. Only they and the intermediaries from whom they have obtained material know whose exuviae is held by a fight magic man; the following must remain, thus, no more than a suggestion.

Among the Tsembaga, all men who are in possession of fight packages happen to be shamans as well (although this need not be the case). I was informed by one of these men that those enemies whom the smoke woman marked for killing through him were among those whose exuviae, or whose father's exuviae, were contained in his fight package. If it is the case that designations of the smoke woman are to some extent self-fulfilling prophesies, then those whose exuviae are contained within the fight package are more likely to be killed than other members of the enemy force.

The declared motive of a man's betrayal of another to his enemy by giving them his exuviae—suspicion of witchcraft or sorcery—is suggestive. While it is entirely possible that the victim of such a betrayal may be the innocent object of unprovoked antagonism, it is also possible that he has departed sufficiently from certain approved modes of behavior to arouse covert, but not general, animosity. Widespread antagonism toward a member of the group is likely to lead to general agreement that he is a witch, and when such agreement exists, betrayal to the enemy is unnecessary; a man's own clan brothers may kill him. No such killings took place among the Tsembaga during the period of fieldwork, but grumblings frequently were heard concerning who would have been killed for a witch had the government not arrived, and descriptions of the personalities of individuals killed for witchcraft in the past indicate that the victims are likely to be bad-tempered, argumentative, and assertive.

Information is both insufficient and general, but it does suggest that those likely to incur covert but less than general suspicion of witchcraft include those who have plenty of women, valuables, pigs, and crops, but who are not generous. For reasons they do not make clear, the Maring regard especially marked success in pig husbandry or gardening to be associated with witchcraft or sorcery. If a man's pigs and gardens prosper out of proportion to those of his neighbors, he invites suspicion. It is the exuviae of such a man, it is suggested here, that is likely to be sent to the enemy, and he is thus expected to be killed in battle. There are strong reasons then for a successful man to be sufficiently generous to allay the envy of those around him. At the least, fear that through treachery one's exuviae might come to rest in the fight package of an enemy, like fear of sorcery and witchcraft and fear of being suspected of sorcery and witchcraft, is a factor in maintaining the social and economic egalitarianism that characterizes Maring society.[8] The fight packages of the enemy may indeed be worthy of the fear in which they are held, because those whose exuviae are contained in them may be in greater physical danger than their clan brothers.

This suggestion must apply mainly to the practice of shamanism by men possessing a fight package, for it is not clear how the whistling of the pig's head is produced. Taboo men say, of course, that it is the head of the pig that whistles, and other people don't know. Few people, or none at all, watch the taboo man's face during the procedure, and it is possible that he simply puckers his lips and blows. It is also possible that the steam of the oven is in some way made to produce the sound.

It may be the case that warnings, like designations of who among the enemy may be killed, are correct in a greater than chance number of cases. If a shaman could judge who in his clan would be marked for killing by his counterpart in the enemy group, or whom the enemy might try especially hard to kill, he could warn this man.

While any member of a local population may be slain in retaliation for a homicide committed by one of its members, it is sometimes the case that the victim's group marks one of the killer's group as the preferred object for revenge. This may be the killer himself, or one of his sons, although according to Vayda it is the custom among some Maring groups to designate a member of the enemy group who physically resembles the victim. Such a practice carries the ideal of absolute reciprocity to the extreme.

Sometime during the night two *rumbim* are planted in the enclosure shared by the *riŋgi* house and the men's house in which the warriors are assembled. These *rumbim,* which are also, according to some informants, called "*yu miñ rumbim,*" are planted to keep the *miñ* of the men inside the enclosure when, on the follow-

8. Both Wolf (1954:46) and Kluckhohn (1944:67–68) have made similar suggestions concerning the functions of fear and suspicion of witchcraft in other kinds of communities in other parts of the world.

ing day, they themselves leave it for battle. That is to say, the men leave their *miñ*, their life stuff, behind when they go to battle. Tsembaga say that if their *miñ* were exposed to the powerful spells made by enemy fight magic men, they would surely succumb, causing their possessors to fall victim to enemy weapons. The *miñ*, therefore, must be left in a safe place and kept from following their possessors into battle or running off if frightened. The *rumbim* prevent them from leaving the enclosure. Informants say that no Tsembaga knows the spells that must be enunciated over these *rumbim*, and that one old Tuguma man comes to perform this ritual for all the Tsembaga.

During the night the fight magic men paint and bespell a number of small stakes. Before dawn they go quietly to the fight ground and push them into the earth far enough so that they cannot be seen. These stakes are said by informants to cause the enemy to lose their *miñ* if they are brought to the battlefield. Informants do not report any encounter of opposing fight magic men on the fight ground. Whether there were conventions preventing such meetings was not investigated.

At dawn the pigs are taken from the ovens. The fat belly of the "leg pig," the pig dedicated to the spirits of the low ground, is covered with salt. When the allies arrive, as soon as possible after dawn, they are given portions of this salted pig fat, which they eat immediately along with the green vegetables, mostly ferns, with which the pig was cooked. The remainder of the pig will be recooked and given to them in the evening, after the day's fighting.

The procedure involving the pig offered to the red spirits, the "head pig," is more elaborate. Before its salted belly is eaten by the principal combatants, they blacken their bodies with the *riŋgi*. The heart and lungs of the pig are placed on a small table in the *riŋgi* house, and the fighting stones are removed from their net bag and laid on top of them. One informant says that the red spirits have themselves come into the stones through the fire. Placing the stones on top of the heart and lungs permits the red spirits to partake of these parts. The men then enter the house two or three at a time to take the *riŋgi*, the black ash made in the fire ignited by the red spirits. They are asked if they prefer to take *riŋgi* from the male fighting stone, the *wai*, which is the pestle, or from the *mai*, meaning in this instance the female, which is the mortar. Their choice has a bearing upon the permanence of some of the taboos, to be discussed below, which are assumed at this time. If the *mai* is selected, the taboo man uses it to apply a little of the *riŋgi* to the shoulders and forehead of the warrior. If the *wai* is chosen, the taboo man applies some of the ash to it and the warrior licks it off. It is unclear whether it is the contact with the stones, the reception of the *riŋgi*, or both which brings the red spirits into the head, but informants agree that this is the effect of the procedure. When men go to battle, their *miñ* are left behind, but the hot, dry, hard red spirits are thought to have come literally into their heads where they burn, informants say, like fires, imbuing them with strength, anger, and the desire for revenge.

After having received the *riŋgi* ritually from the taboo man, the warrior retires from the house with a handful of ash so that he may blacken his entire body. Now that the red spirits are inside him and he himself is "hot," he may apply the "hot" protective and strengthening ash to his skin. Had he done this before being marked by the female stone or licking the male stone, informants say, his skin would have come up in blisters from the supernatural heat of the *riŋgi*.

Riŋgi is also applied, Tsembaga say, so that the face of the warriors may not be seen by the enemy, in accordance with one of the taboos assumed when the fighting stones were hung a few hours earlier. It is said that if the enemy can see a man's face, they will be able to kill him.

The taboos assumed with application of the *riŋgi* are extensive and include both limitations on social intercourse and prohibitions against certain foods. The taboos on social intercourse refer mainly to relations with women. A man cannot, while he is wearing *riŋgi*, eat food from the same fire as a woman or cooked by a woman. While he may speak to women, he may not touch them; sexual intercourse is, of course, forbidden. Women, and anything touched by women, are said to be "cold." Contact with them would extinguish the fires burning in the men's heads. Conversely, contact with the hot men would, it is said, literally blister the skins of the women.

Food taboos are extensive. All wild animal foods are forbidden. Snakes, eels, catfish, lizards, and frogs are proscribed because they are "cold." Marsupials are not eaten because they are the pigs of the red spirits and cannot now be taken until the living are in a position to give their masters pork in return for them. Some green vegetables are also forbidden. These include *Setaria palmaefolia* and hibiscus leaves. The latter are forbidden because they are wet and slippery; their ingestion will not only extinguish the fires in the head but also make it difficult to grip an enemy in hand-to-hand combat. No explanation for the taboo against *Setaria* could be obtained, and it may be that it was merely imported with the plant itself or else developed in a former home area long ago. Taboos against the ingestion of this plant by males are in effect elsewhere in the highlands (see Newman, 1964:263–66). Some bananas and yams are also tabooed. These are, in the main, soft varieties, which it is said would injure the hardness or strength of their consumers. All food grown in the *wora* must be avoided, and unless the fight ground is located there men must not enter the *wora*.

All of these taboos remain in effect until, at the earliest, the *riŋgi* is ritually removed at the end of the fight. If a man has taken *riŋgi* from the male stone some taboos remain for the rest of his life or at least until he is a very old man. These permanent taboos include prohibitions against the ingestion of snakes, eels, catfish, lizards, frogs, and many of the marsupials that live in the high forest. Since most men take *riŋgi* from the male stone the great bulk of these subsidiary protein sources are, as it has already been mentioned in an earlier chapter, reserved for women and children.

Permanent prohibitions also apply to certain vegetables. These include all varieties of *Setaria palmaefolia*, some yams, and some bananas. There are, in addition, a number of other plants which may be eaten, but never from the same plant that has fed a woman. Several varieties of *Pandanus conoideus* fruit fall into this category, as do sugar cane, banana, and *Saccharum edule* varieties that have red skins or leaves.

After the *riŋgi* has been applied, a hasty meal is made of the greens with which the pig has been cooked. The pork, however, except for the bellies, is returned to the ovens for recooking with fresh greens, in preparation for the meal at the end of the day. According to information I received, the men then form a circle and consume fighting leaves and heavily salted belly fat. Georgeda

Bick (personal communication) received slightly different information from some of the same informants in 1966. She was told that the salt was not taken on pig belly, but wrapped in the leaf of the succulent *komerik* (?*Pollia* sp.). Informants say that eating the fighting leaf makes them wild and bloodthirsty, but the effect is probably psychological rather than physiological, for the ingestion of the prescribed amount produced no effect upon me. No explanation involving specific supernatural or natural processes could be elicited about the consumption of pig belly. Bick, however, was told that the ingestion of salt and *komerik* would ward off enemy spears.

Whether the salt was ingested with pig or with *komerik,* the practice may have some important consequences. First, the consumption of salt is accompanied by an absolute taboo upon drinking any liquid whatsoever while on the fight ground. Certainly no man who wears *riŋgi,* and perhaps no allied warrior, although this is not clear, may drink water or the juice of sugar cane, or eat cucumbers or papaya while the engagement continues. The rationale for this taboo, in light of previous discussion, is apparent. These cold, wet foods will extinguish the fire that burn in a man's head, leaving him weak and defenseless. A man, therefore, does not drink from the time that he consumes the salted pig belly until he leaves the fight ground later in the day. To note that the resulting thirst might well limit the length of the fighting day is not to be facetious, particularly in light of the fact that fighting takes place usually on sunny days. (Fighting is postponed if there is rain, for rain would extinguish the fires in the men's heads.) Indeed, some Tsembaga men have admitted that fighting would cease for the day when the warriors became too thirsty to fight any longer. The ingestion of salted pig belly and the associated taboo on taking fluids may thus mitigate the consequences of combat by shortening the fighting day. Other consequences of salt and fat consumption, however, may operate to increase the intensity of the fighting while it does last.

The Maring diet is deficient in sodium. This is obvious because most Maring will eat pure table salt by the handful, and sodium appetite appears to be directly correlated with sodium need (J. Sabine, personal communication, 1965). The manufacture of native salt was discontinued some years before fieldwork began and no samples could be obtained. Because the native salt was made from the waters of mineral springs and not reduced from plant ash, however, it is quite certain that it was a sodium salt, despite the lack of samples. Eating large amounts of sodium just before fighting would have permitted the warriors to sweat normally while maintaining normal blood volume. If they lacked this additional sodium they might well have experienced a lowering in blood volume and a consequent weakness as they lost sodium through sweating (W. V. Macfarlane, Frederick L. Dunn, personal communications, 1965).

Sabine (personal communication, 1965) points out that the salt may also play a part in the catabolization of the fat, which, according to Macfarlane (personal communication, 1965), would then become available as energy about two hours after consumption, thus providing a "second wind" to the tiring warrior.

The consumption of the lean pork at the end of the day, it may be mentioned, would offset to some extent the nitrogen loss caused by the stress of fighting.

After eating the fighting leaves and salted belly, the men form a circle and fight magic men press the fight packages to their hearts and their heads, to the cassowary plumes adorning their heads, and to the weapons as well. Another fight magic man applies bespelled *gir*, a gray clay, to their ankles and feet. The taboo men rub *riŋgi* on all of the jabbing spears and axes.

Having finished applying the fighting package to men and weapons, the fight magic men build a fire outside of the circle formed by the men and lay lengths of green bamboo on it. Both sets of ancestors are asked again for their aid and protection.

In the meantime, the men have enclosed the inside of the circle by making a fence of their shields. The taboo men are inside this circle, and when the bamboo explodes they begin to scream out to the red spirits, asking again who may be killed among the enemy and who among the locals are in great danger. Informants say that this is the only occasion upon which the red spirits themselves communicate with the living without going through the intermediary smoke woman. The taboo man screams, leaps in the air, hurls himself against the wall of shields. His eyes roll up into his head showing only the whites and his tongue protrudes. He screams out the names and, after writhing on the ground, lies still for a moment or two. He is assisted to his feet by one of the fight magic men while the men yell war cries. More bamboo is exploded and the men, led by a fight magic man waving his fighting package in the direction of the fight ground, leave for battle brandishing their weapons, singing a song called "*de*," and shouting war cries. The taboo men, who recover quickly, join the procession.

In contrast to the weapons of "small" or "nothing" fight, both axes and two types of jabbing spear are employed in the "ax" fight. The point of one type is armed with a sharpened beak of a hornbill, which remains in the wound when the weapon is withdrawn. The other is barbed for half of its seven-to-nine-foot length. Below the barbs three or four prongs project forward and outward at an angle to the shaft. Such a weapon seems designed for defense against headlong charges, which informants say were not a regular feature of Maring warfare.

On the fight ground formations were mixed; both principal combatants and allies stood side by side, it is reported. The taboo man remained in the rear, heavily protected by the men closest to him. The slightest wound sustained by a taboo man supposedly portended the inevitable death of one of the men of his group, and if he were killed it meant that his side would be routed. Such a belief could operate, of course, to mitigate the severity of fighting if both sides spent their energy in trying to shoot the taboo men of their opponents and if his men fled when he was killed. There is no suggestion from any informant, however, that ax fights degenerated into attempts to shoot the enemy taboo man while others of the enemy were being ignored.

There is little information concerning leadership on the fight ground. It is quite clear, however, that discipline was not tight nor control close. Formations were several ranks deep, and the men in the front row were relieved by those behind them from time to time. While the opposing men in the front ranks were fighting a series of duels from behind their enormous shields, bowmen in the ranks farther to the rear gave support to the men in the first rank and shot at those who exposed themselves. Fatalities generally occurred when a man in the front rank was brought down by an arrow. His opponent, supported by nearby comrades, would then rush in to finish him off with axes. The fallen man's comrades would also rush to his defense, protecting him with their shields if possible.

Informants' accounts give the impression that this static positional fight was most common. In most engagements the opposing forces stood toe-to-toe behind their huge shields with no tactical or strategic movement taking place. A series of such engagements frequently continued sporadically for weeks, with neither side suffering more than light casualties. The antagonists, it seems, were waiting for the day when their enemies would arrive on the fight ground without the support of their allies. On the day that a group found itself with a clear numerical advantage, instead of taking up the usual static positions it would charge.

Maintaining the support of allies seems to have become increasingly difficult as the fighting went on. Informants say that they simply tired of participating in a fight that was not their own. Victory, thus, must have often gone to the group who could continue to mobilize its allies the longest. There are two clear-cut instances of this involving the Tuguma and the Tsembaga. In their last conflict with the Kundagai, it was on the day that their allies, the Tuguma, did not appear that the Tsembaga were routed. The Kundagai, heavily supported by the Kanump-Kaur and other groups, took advantage of the absence of the Tuguma to mount a charge. The Tsembaga suffered eighteen fatalities on this one day, six of them being women and children, and many other people were wounded. Thus casualties sometimes were heavy.

The Kanump-Kaur avoided the more dire consequences of the failure of their allies to appear for an engagement against the Tuguma. When they heard that the Kandembent-Namikai, their most important supporters, were not going to arrive, they themselves didn't go to the fight ground. They simply gathered up their women, children, and pigs and fled their territory. Vayda has gathered accounts of similar behavior among a number of Jimi Valley Maring groups.

Waiting for a numerical advantage and then charging, according to informants' accounts, was the most frequently employed strategic and tactical procedure. There were others as well, however. An example of ambush is to be found in Tuguma history. Night-time raids were rare but did occur. Accounts of battles indicate, furthermore, that flanking and surrounding movements were sometimes attempted; by stealth a force would gain its enemy's rear or flank, opening him to either cross fire or charges from two directions. These latter maneuvers were probably only attempted by forces with some numerical superiority.

Informants' accounts indicate that an ax fight often went on for weeks, or perhaps even months; however, engagements did not take place every day. Rain confined the men to their houses, and the requirements of ritual also produced interruptions. The frequency of ritual performance was, to a large extent, directly correlated with the frequency of casualties, for most of the rituals were performed in connection with casualties.

After a fatality, fighting was discontinued while the group that suffered the loss mourned the deceased and performed the necessary funerary rites, which included killing one or more pigs. Those who inflicted the fatality also had to undertake ritual performances to protect themselves from the spirit of their victim. It is said that the *miñ* of the slain man follows the killer home, menacing not only the killer himself but all of his coresidents. If the killer sleeps in the men's house the spirit of the slain man might follow him inside, wreaking vengeance upon all who are present. When a man kills an enemy, therefore, he sleeps away from his fellow warriors to keep from exposing them to danger. To protect himself, he sleeps in the *riŋgi* house.

In the morning the killer slaughters one of his own pigs for the red spirits. In his address he tells them that with their help he has killed an enemy and he now gives them a pig in thanks for their assistance. He asks that the *miñ* of the slain be prevented from killing him or any of his brothers. He invites the red spirits first to eat the pig being offered them and later to assist the living in killing more of the enemy. Since the allies may not partake of pig offered to the red spirits, a second pig may be killed and dedicated to the spirits of the low ground, but it is not clear that this is the case.

While the pig is in the oven the fight magic men prepare to drive the spirit of the slain man from the territory through a procedure called "extraction and disposal magic" (*kunda guio warumbon*). They fashion brooms from *rumbim* and obtain from the forest several 4- to 6-foot lngths of the vine *deraka* (*Pipturus* sp.), leaves of *močam* (*Aglaomena* or *Alocasia* sp.), and a few 3- to 5-foot poles that are 1 to 2 inches in diameter of *ganč*, a tree of the *Rubiaceae*

View north across the Simbai Valley to the gardens of the Kekai (light areas) from above the Tsembaga dance ground. The woman's house in the foreground was later demolished to enlarge the dance ground.

The author, who is 5′10″, with a group of Maring men and boys.

Large pigs being led to slaughter.

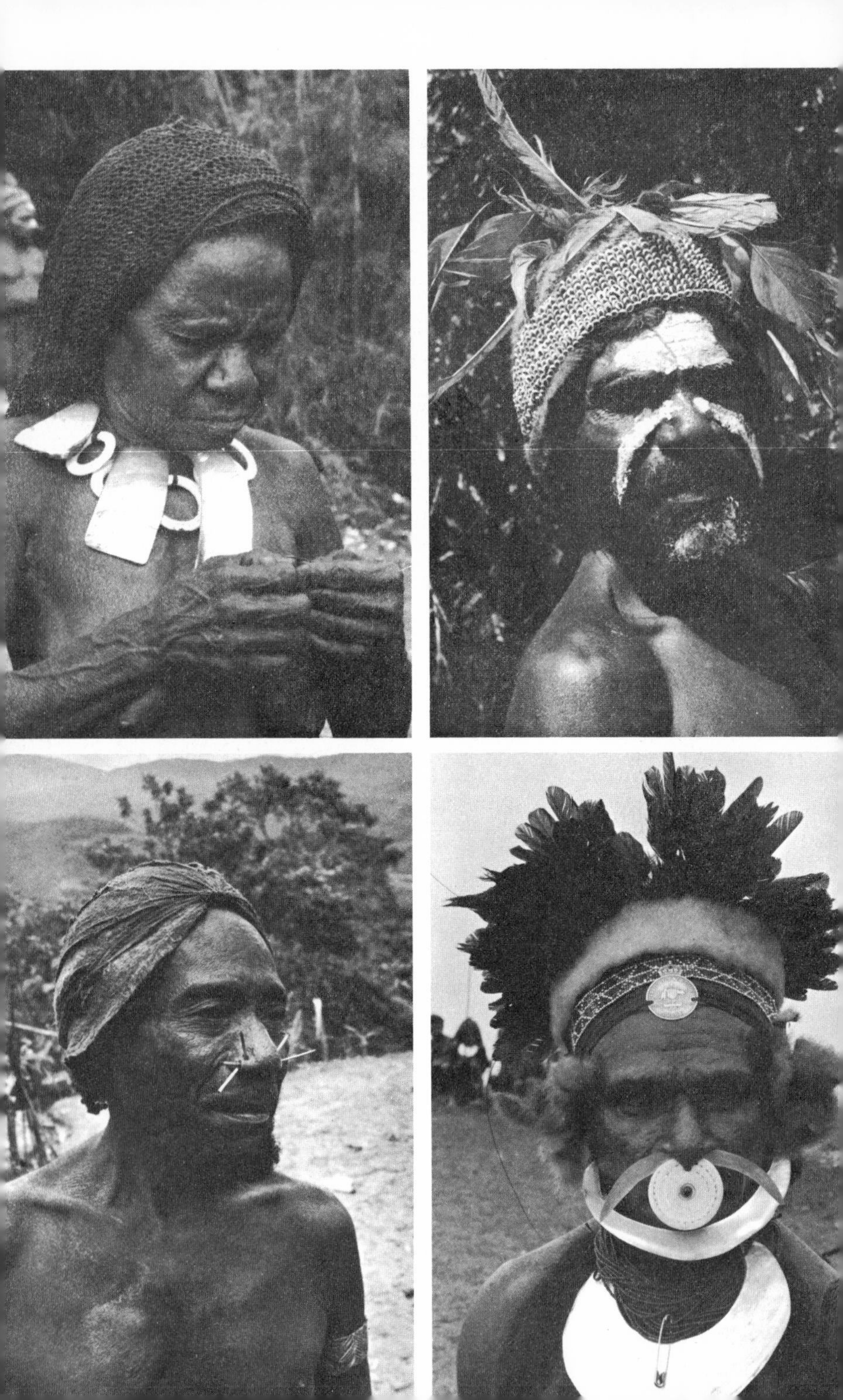

Joints of a wild pig are distributed by a successful hunter. Every Tsembaga shared in the flesh of this animal.

Members of the Tomegai and Kwibigai clans join to feast on the wild pig in a men's house enclosure.

Bespelling sections of bamboo before their use in a stake planting ritual. The stakes (left foreground) and the močam leaves (behind the bamboo) are also treated.

Tsembaga men ready to leave for Tuguma to participate in the Tuguma stake planting ritual. When the bamboo being placed on the fire at the right explodes, they will depart.

(*Opposite*)

The trail crosses a boundary between local populations. It is rare for shields to be set at boundaries, but stakes (left foreground) are always planted.

Erecting the *timbi* house. Bark is scraped from the center post so that it will fall on the wealth objects, causing them to increase.

(*Opposite*)
A man leaps onto heated cooking stones to pierce the *marita* fruit (in his left hand) with a cassowary bone.

A medium-sized pig is sacrificed to the Red Spirits. (Photo taken by Mrs. Cherry Vayda among the Fuŋgai-Korama, a Maring group living east of the Tsembaga in the Simbai Valley.)

(*Opposite*)
A woman renounces a taboo.

The Auŋdagai charge onto the Tsembaga dance ground.

Men bedecked for dancing. The young man in the center wears a *mamp gunč*.

Trade follows dancing. Two men discuss the merits of a gold-lip shell.

family. The posts are sharpened on the bottom and painted at the top with earth pigments. Spells are said over these four kinds of objects and they are passed out among the men. Bamboo is laid in a fire and the ancestors are asked to assist the living in driving away the spirit of the slain man. When the bamboo explodes the men stamp their feet and cry out. Those who are carrying the bespelled objects run all over the enclosure. The men with *rumbim* make sweeping motions as they run. Those carrying the vine *deraka* whip the ground, while the men with the sharpened *ganč* posts thrust their points into the earth at frequent intervals. The *močam* are carried by their 2- to 3-foot stems with the leaf held a foot or so above the ground. As the men run they occasionally shout, *"Pio!"* ("Go!").

After having covered the enclosure thoroughly, they go out through the gate and head for the houses of the women. Other men follow, brandishing weapons and shouting battle cries. Every house and every enclosure is visited so that the spirit of the slain man, wherever it is, and the supernatural corruption (*tukump*) it has spread, may be beaten by the *deraka*, pierced by the *ganč* posts, "bitten" by the *močam* leaves (which contain oxalic acid), and swept by the *rumbim* toward the boundary. There, amidst battle cries, the four plants are deposited, after spells are again recited over them to bring harm to the enemy by sending to their territory not only the spirit of the dead man, but also his *tukump*.

While neither rituals performed by the victim's nor the killer's group require more than a day or two to complete, fighting may be discontinued for considerably longer; a period of five to seven days, informants say, was not unusual. It is the victim's group that calls out the challenge for a new engagement but sometimes they delayed, since fatalities, it is admitted, do damage to the faith of the victim's group in the efficacy of its ritual performances. After a death, therefore, requests might be sent to renowned fight magic men in friendly groups, asking them for their assistance. By the time that the invitations reached such men and they had traveled to the territory of those who had sent for them, several days might have passed. These "cease-fire" periods were used by combatants on both sides to take care of garden work if necessary. Another day might then be spent in preparation for the visitor's rituals and in their performance. After their completion the challenge for a new round of fighting could be issued. Both sides again killed pigs for the red spirits and the spirits of the low ground and sent word to their allies that the time for a new engagement had been set.

Wounding, when it was serious, also required interruptions of the fighting, according to informants. The victim's group would on these occasions kill a pig for the red spirits, which would be eaten by the victim and his agnates. New engagements might be delayed until it was clear whether the injured man would live or die.

Several obvious advantages of "ritual cease-fires" may be summarized here. First, interruptions in the fighting provided the combatants with an opportunity to perform necessary subsistence tasks. Second, they relieved the psychological and physiological stress that must have accompanied an ax fight. The requirement of the rituals for the slaughter and consumption of pigs, furthermore, may also have served to mitigate the ill effects of the increased catabolization of protein brought about by this stress. Third, it is possible that such interruptions may have enhanced the opportunities for reestablishing peace on a more prolonged basis. For one thing, each cease-fire resulted in a partial or full demobilization of the allies, who returned to their own territories. Informants' accounts suggest that remobilizing allies became increasingly difficult as fights dragged on. A group having difficulty rallying its supporters for a new encounter might choose not to issue the challenge, or, if the homicide score was not too uneven, to send word to the enemy, either through intermediaries or by shouting, that a truce was desired. This was likely to be acceptable if the fighting had been protracted and some casualties had been suffered. Informants report that men would say that enough had been killed or wounded for the present, that if the fighting continued everyone would be killed, and that it was better to perform the truce rituals now and to resume the fighting again sometime in the future.

TERMINATING THE FIGHT

Most ax fights ended in a truce. Sometimes, as suggested above, termination of the hostilities was arranged during periods when fighting had already been interrupted by the necessity to perform rituals. At other times it was arranged while both sides were on the fight ground.

Informants say that such agreements were reached at least three times by the Tsembaga and their enemies the Kundagai. Evening the homicide score, while not necessary, no doubt facilitated the invocation of truce. Old informants say that in the first fight against the Kundagai, when the score was evened by the slaying of a Tsembaga man (each side had then suffered four deaths), the Tsembaga called out to the Kundagai that their requirements for revenge had now been fulfilled and that both sides should return to their respective houses to plant the *yu miñ rumbim,* signifying truce. On the other two occasions when truces were declared, there were disparities in the homicide score but the contestants agreed that killing had been sufficient for the time being and that the pressures of gardening and pig herding required a cessation of the fighting.

Truces continued for the duration of the ritual cycle. The description and timing of this cycle will be discussed in the next chapter; it is sufficient to note here that these ritual cycles were protracted, as were the truces that coincided with them. They sometimes endured for fifteen or even twenty years, although ten or twelve years was probably the average.

Not all fights ended in truces, however; it sometimes happened that one side was routed. The victors, in such instances, after killing every man, woman, and child unfortunate enough to be caught, laid waste the territory of the vanquished. After the defeat of the Tsembaga in 1953, the Kundagai tore up the Tsembaga gardens, cut down most of the *pandanus* and *ambiam* (*Gnetum gnemon*) groves, desecrated the *raku,* the burial places where pig sacrifices are made, burned the houses, and killed all the adolescent and adult pigs. Their depredations were directed toward making it as difficult as possible for the Tsembaga to return to their own territory, rather than toward the acquisition of booty. Since the Tsembaga were *čenaŋ yu,* or formal enemies, to them, foodstuffs grown by the Tsembaga were supernaturally proscribed. Their booty was therefore limited to baby pigs, which could be taken home to be raised, and perhaps to wealth objects, such as plumes and shells, which the fleeing Tsembaga had left behind. The slain pigs were not allowed to rot, however. Neutrals, seeing the smoke of the burning Tsembaga houses, immediately came and carried the carcasses back to their own houses for cooking.

Having done what they could to lay waste Tsembaga territory, the Kundagai retired to their own ground. The immediate occupation of Tsembaga territory was tabooed to the Kundagai. It will be recalled that when the fighting stones are hung up, a taboo against entering enemy territory except in anger comes into effect. Once the armed conflict had come to an end and despoliation had been completed, the Kundagai not only returned to their own ground, therefore, but were restrained from reentering Tsembaga territory by fear of both the ancestral spirits of the Tsembaga, who were thought to remain at least for a time on the ground lost by their living descendants, and also by fear of their own ancestral spirits from whom the injunction emanated.

If the defeated do not return to their territory, it remains vacant while the fighting stones of the victors remain hanging from the center poles of their *riŋgi* houses. I know of only one instance in Maring history in which the territory of the routed group was immediately occupied by the victors. In 1955 the Kauwasi, a Jimi Valley group, routed their neighbors, the Cenda. Some members of the victorious forces announced their intention to make gardens immediately on the ground abandoned by their opponents, but they were actually prevented from doing so by the Australian government, whose presence was first being felt in the area.

This case may illustrate the limits beyond which a particular set of supernatural constraints cannot contain demographic and ecological pressures. The Kauwasi are the most numerous of all Maring local populations. They numbered, in 1963, nearly 900 persons, according to Vayda, and there is no reason to believe that they were fewer in 1955. In 1955, moreover, their numbers had been swollen by refugees from the Tsembaga, Ambrakwi, and Tsengamp. There must have been at the time over 1,000 persons living on Kauwasi territory, and 1,100 is not unlikely. The land of the Kauwasi, moreover, is poor. Large areas are in grass, and much of the gardening takes place in very young secondary growth. There are indications, in short, that the Kauwasi were pressing the carrying capacity of their territory, a state of affairs that could only have been worsened by the influx of refugees.

It is further interesting to note that some but not all Kauwasi informants deny any intention of occupying the territory of the Čenda themselves. They say that it was not the Kauwasi but the

Tsembaga, Tsengamp, and Ambrakwi refugees who were going to take Čenda land, and that they could do so because they were not themselves enemies of the Čenda but were merely assisting the Kauwasi in the fight as allies. This statement is inconsistent with other statements which indicate that the refugees were considered principal combatants, and not allies.

Accounts of routs suggest that in some instances the members of the vanquished group do not return to their territory. The group fragments as the survivors flee to affines or cognates. The Tsembaga who survived the 1953 disaster took refuge with seven different local populations, two of which were across the Simbai River to the north, while three were to the south, across the mountain in the Jimi Valley. To rally these small and scattered remnants for an attempt at reoccupation would have been difficult in light of their dispersion, their abiding fear of the Kundagai, and their ruined *pandanus* and *ambiam* groves. The Tsembaga ceased to exist as a group after their defeat, and, if it were not for agents of the newly arrived Australian government who offered to protect them, it is unlikely that they would as a group have returned to their territory. Not only was their return discouraged by the dangers and difficulties of such an enterprise, but continued residence with their hosts was encouraged by participation in rituals that had the effect of converting their de facto membership in these local populations to de jure membership (this will be discussed in the next chapter).

While accounts indicate that the return of an entire local population to its abandoned territory often did not occur, frequently, and perhaps usually, a fraction of that territory would be reoccupied by remnants of the vanquished group under the cover of one or more of the groups among whom they had taken refuge. Some Tsembaga, for example, after being driven off their territory by the Kundagai, took refuge with the Tuguma immediately to the east. As members of the Tuguma group, or under their protection, they reentered the eastern portion of their territory and made a few gardens. They were accompanied in this enterprise by a few Tuguma who had been told by Tsembaga relatives living elsewhere to avail themselves of their lost ground. So far as I know, the reoccupied Tsembaga territory was used for gardening only; houses remained at a greater distance from the enemy border. Such limited reoccupation of abandoned territory is reported in two other instances in Tuguma-Tsembaga history.

THE TRUCE RITUAL: PLANTING THE RUMBIM

Even when truces have been arranged on the battlefield, the antagonists remain uncertain of each other's sincerity. If after a protracted period active hostilities are not resumed, truce-making rituals are performed. Informants, who it must be admitted are inaccurate in such matters, indicate the period is a month or so. In addition to a sufficient time lapse since the last engagement, it may be that word certifying the resolve of their opponents to terminate the fighting comes to each of the principal combatant groups through neutrals. In light of the reticulate nature of intergroup relationships throughout the Maring area this is likely, but no statements from informants give support to the possibility.

Truce rituals are performed only by groups that remain on their own territory after the fight. The scattered survivors of a rout simply remove the *riŋgi* from their bodies with little ceremony. The antagonists do not participate together in truce rituals; each conducts them separately. When a group is assured that its opponents have given up any intention of prolonging the hostilities, word is sent to the allies to assemble for the killing of the pigs and the planting of *rumbim.*

The number of pigs slaughtered depends upon the size of the herd, the length of the conflict, and the number of casualties. The sacrifices take place at the *raku,* where ancestors' houses had been located. These are scattered throughout the territory, and every man has a number at which he sometimes sacrifices pigs. The specific *raku* at which slaughter is to take place are designated on this, as on all other occasions by shamans, who tell each man where the ancestors wish to eat their pork. On pig-slaughtering occasions during the period of fieldwork shamans' advice directed most of the members of a subclan to the same *raku,* but there were always some men killing all or some of their pigs by themselves in isolated *raku.* Informants say that this was also the case when *rumbim* was planted. The locals are accompanied to the *raku* by their affines, cognates, and, perhaps, nonkin trading partners from other places who assisted them in the fight.

Some of the pigs killed are for the red spirits, but most are offered to the spirits of the low ground. The burden of the addresses to both categories of ancestors is to thank them for their assistance in the fight. Most of the pork from animals dedicated to the spirits of the low ground is immediately presented to the allies, who take it home with them when the rituals are concluded. The fat bellies are first removed, however, for formal presentation. This belly fat, along with that taken from the pigs dedicated to the red spirits, is brought to the men's enclosure, where the *riŋgi* house stands. This is the scene, on the following day, of the planting of the *rumbim*.

Everyone, including allies, women, and children, assembles for this ritual. In preparation all varieties of wild animals are captured: marsupials, snakes, lizards, frogs, rats, insects, grubs, and birds. A special oven (*pubit*) is prepared out of bark. It is about three feet square and rests directly on the ground. This oven is loaded with greens, the wild animals, and the belly fat of the pigs dedicated to the red spirits, which is cut into little pieces. While this is cooking the presentation of the belly fat of the pigs dedicated to the spirits of the low ground is made to the allies. These men sit in a circle and, as each man's name is called his local affines, cognates, or trading partners push handfuls of the now cold salted belly fat into his mouth. A second local relative assists by pushing additional salt into his mouth and by sprinkling more on his shoulders. There is a vague order of precedence in these proceedings: the sons of men slain in the fighting, if there are any, will be called first, and men who have killed enemies next.

After the presentation to their allies has been made, the principal combatants may remove their *riŋgi*. The fight magic men bespell the cloth-like tents of certain tree caterpillars, which are used as wash cloths, and the sap taken from the tree *yiŋgam* (unident), which serves as a soap.

These spells are among those included in the category of *kunda*, known as "turning word" (*andik meŋ*). Throughout the remainder of the entire ritual cycle, *andik meŋ* is performed from time to time to "turn off," or reverse, magic performed or the ritual states assumed in connection with fighting.

As the oven is opened the taboo man addresses both categories of ancestors, recounting the circumstances of the fight and announcing to them that the taboos associated with the *riŋgi* will now be removed. The men then cleanse themselves of *riŋgi*. The removal of the *riŋgi* signals the end of the taboos against sexual intercourse, food cooked by women, and some of the foods forbidden during the time of fighting. The men, along with the women and children, may now eat the food in the oven. Informants say, when questioned about it directly, that those men who took *riŋgi* orally from the *wai*, or pestle, avoid the snake, lizard, rat, and other forbidden foods that the oven contains, while those who received *riŋgi* from the mortar, or *mai*, eat everything, as do the women and children. This may not be the case, however. It must be hard to distinguish a small piece of forbidden cooked rat from a small piece of perhaps acceptable marsupial, and it must be impossible to distinguish pieces of acceptable marsupial from marsupials that are tabooed. This may be, in short, an occasion when food taboos are in fact abrogated simply because mistakes may be impossible to avoid.

After the contents of the oven are consumed, the *yu miñ rumbim*, always a red-leafed variety of *Cordyline fruticosa*, called "*tundoko*," is planted. After it has been bespelled with *andik meŋ*, it is placed directly in the middle of the square bark oven in which the food has just been cooked. The ancestors are addressed by one of the fight magic men, who thanks them for both preserving the lives of the men who survived the fight and for not allowing them to be dispossessed of their territory. He repeats that the fight is over and tells them that the *rumbim* is now being planted on their own territory. He assures them that while it remains in the ground the living will not fight or turn their thoughts to fighting. He deprecates the amount of pigs given to them and to the allies, but promises that while the *rumbim* remains in the ground the thoughts and efforts of the living will be directed only toward raising taro and more pigs so that some time in the future, when the pigs are of sufficient size and number, the *rumbim* may be uprooted and pork in great quantity can be given to them and to the allies for their assistance in the fight.

He also asks the spirits to care well for the *rumbim*, for if it does not flourish neither will the men. All of the local males are then directed to place a hand on the plant as the earth is tamped around its roots. The taboo man tells them that intercourse with their wives will now result in children and that the children will be strong and grow quickly.

Some informants say that the laying of hands on the *rumbim* actually introduces the *miñ,* the life stuff of the men and boys, into the *rumbim,* where they remain for safe-keeping. Others deny that this is the case and explain that while this *rumbim* is planted for the sake of the men, their *miñ* remain either in the *rumbim* planted the night previous to the ax fight or confined by that *rumbim* to the men's house enclosure. It has already been mentioned that both are sometimes referred to as "*yu miñ rumbim*" (men's "souls" *rumbim*), but the exact whereabouts of the men's *miñ* is not a matter of great concern here. It is clear, however, that the *rumbim* is associated with males; in fact, women may not touch this plant.

Another plant, *amame* (*Coelus sentellariodes*), a low, fragrant, herbaceous, in this case green-leafed, ornamental, is planted for women. When the men unclasp the *rumbim,* the taboo man plants the *amame* around the outside of the oven. Some informants say that small pieces of fat from the belly of the pig are planted with the *amame,* and all agree that it is referred to as the "pig belly *amame*" (*konj kump amame*). While planting it the taboo man asks that the spirits of the low ground care well for the *amame,* that the pigs grow fat, that the women be fertile, and that gardens flourish.

The sexual and fertility content of this ritual seems apparent. The spatial placement of the ritual objects associated with male and female may represent the procreative act, which in turn symbolizes fertility in general. It is in the receptacle that later received the *rumbim* that the varied fruits of the land were cooked, and pigs and gardens are associated with the same plant as are women. It must be mentioned here, however, that *rumbim* and *amame* are also planted over graves, and death as well as fertility might also be symbolized here. It has already been noted, however, that death and fertility are not antithetical in the Maring view.

It seems apparent too that this ritual, following as it does the removal of the *riŋgi,* which terminates the supernatural restrictions associated with fighting, reorients the participants toward peaceful activities. Their attention and efforts are directed away from their borders and those who live across their borders. They may now turn inward and devote themselves to prospering within their successfully defended territory. The cleansing of the *riŋgi from their bodies*, followed by the planting of the *rumbim* and *amame*, ends war and begins peace.

Another important aspect of the ritual planting of *rumbim* is its affirmation of the connection of the individual to the group, and the group to its territory. The ability to plant *rumbim* indicates the successful defense of the territory against outside threat. *Rumbim* may be planted only on the territory. Groups who have been driven off their land cannot as groups perform this ritual elsewhere. Survivors of routs, after having taken *riŋgi* from the fighting stones of their hosts during their hosts' fights, participate in planting *rumbim* with their hosts. This act attaches refugees to the land of the groups who have harbored them, and, moreover, clears the way for the annexation of their abandoned territory by those who drove them off it. When a refugee sacrifices a pig in connection with the *rumbim* planting of his hosts, he invites his ancestors to leave their old residence and partake of pork in the new place. Thus, even the ancestors of a routed group eventually vacate the territory from which their living descendants have been driven.

The peace that follows the planting of *rumbim* is a truce only. Men say that it is still the "time of the fighting stones." These objects remain hanging from the centerpole of the *riŋgi* house, and intercourse with the enemy remains forbidden. Moreover, the time of the fighting stones, that is to say, while the stones remain suspended from the centerpost and the *rumbim* remains in the ground, is a period of indebtedness, for the allies and ancestors have not yet been fully repaid for their assistance in the fight that has just been concluded.

Several taboos remain in effect during this time. Eels, which are said to be the pigs of *koipa maŋgiaŋ*, may not be eaten, and marsupials, which are said to be the pigs of the red spirits, may not be trapped. Before the living may again avail themselves of these "pigs" raised by the spirits it is necessary for them to raise pigs themselves to give in return.

It has already been mentioned that pigs killed in connection with the pig festivals of other groups may not be eaten by men. Informants say that men would be ashamed to eat the festival pig of other people when they have not yet raised sufficient pigs to conduct a festival themselves, and such a pig, when it is received, is given to the women and children.

A taboo on beating drums also prevails. While their *rumbim* remains in the ground the members of a group, when invited to dance elsewhere, dance with weapons rather than drums in their hands.

Further taboos remain that seem to segregate some of the components of the community associated with the red spirits (and thus warfare) from some of those associated with the spirits of the low ground (and thus peaceful activities). Marsupials may not be trapped; although they may be eaten, they may not be mixed with *pandanus* fruit, which is associated with the spirits of the low ground. Red-skinned sugars, bananas, and *Saccharum edule,* as well as certain varieties of *pandanus* and certain species of marsupials, may be eaten by both men and women, but men and women may not share the same plant or animal. Men, moreover, may not eat certain soft tubers, particularly some varieties of *Dioscorea alata, D. bulbifera,* and *D. esculenta,* which grow in the *wora.*

Two of these taboos have certain empirical effects. First, as was mentioned in the last chapter, the taboo against the consumption of "festival pig" from other places applies only to adult and adolescent males. Women and children, who are most in need of protein, are the physiological beneficiaries of this taboo. Second, the taboo against trapping marsupials can only redound to the advantage of the marsupial population. During the period of fieldwork, although no count was kept, it was clear that almost all marsupials taken were taken in traps, rather than by shooting. The yields of trapping, moreover, seemed substantial. When trapping is permitted, several hundred animals may be taken in the course of a year. There are no data concerning the population dynamics of marsupials inhabiting Tsembaga territory, but it may be that the prolonged prohibition on trapping permits these animals to recover from exploitation that might otherwise decimate them.

While these taboos may enhance the well-being of the local population, they are unimportant in comparison to another concomitant of the local group's debtor status, acknowledged with the planting of the *rumbim. The spirits and the allies must be rewarded for their assistance in the recent fight before aggression may again be undertaken.* If a group attacks its enemy before these debts are paid, help will be forthcoming from neither spirits nor allies. Informants agree that both allies and ancestors would be sufficiently angered to refuse to assist a group who initiated new hostilities before fulfilling obligations accrued in the last round of fighting. These sanctions are sufficiently frightening to make breaches of the peace rare.[9] A "truce of God" thus prevails until there are sufficient pigs to present to the allies and spirits and to uproot the *rumbim*. The next chapter will discuss what constitutes "sufficient pigs," how long it takes to acquire them, and the festival during which they are sacrificed.

9. Out of a corpus including information on more than twenty fights, three cases of violations are known. It is interesting that in two of these three cases the chief violators were the Kauwasi, who, it was mentioned earlier, are the largest of the Maring local populations and who may well have been pressing the populational limits of their territory at the time. It is further interesting to note that violation of the *rumbim* truce did not necessarily involve a simple abrogation of the *rumbim* convention, at least in the case of the Tukmenga, a large Jimi Valley local population who joined with the Kauwasi in an attack upon the Monambant several days after the Kauwasi routed the Čenda. According to Vayda, who revisited the Maring area in 1966, the Tukmenga, although sympathetic to the Kauwasi proposal for a concerted attack upon the Monambant, were hesitant because their *rumbim* remained in the ground, and thus they feared an adverse reaction from their ancestral spirits, who had not yet been repaid for their assistance in the last fight. Their solution was to slaughter a small number of pigs, at least some of which were laid on the roof of the *riŋgi* houses as offerings, no doubt, to the red spirits. In other words, the sanctified *principle* that obligations to ancestors must be fulfilled before new hostilities are initiated may have remained unchanged, while the *rule* concerning how those obligations are to be fulfilled was modified. Informants later said, according to Vayda, that the procedure adopted by the Tukmenga had always been acceptable, but no precedents were cited. Data, although insufficient, warrant the suggestion that the Tukmenga were reinterpreting the behavioral requirements of sacred tradition in terms of contemporary conditions, a process well known in Western societies.

CHAPTER 5

The Ritual Cycle

THE DURATION OF THE CYCLE

The length of the ritual cycle, and thus the length of the truce that depends upon the ritual cycle, is largely regulated by the demographic fortunes of the pig population. The *kaiko*, the year-long festival which culminates the ritual cycle, is initiated by uprooting the *rumbim* that was planted after the fight. To uproot the *rumbim* requires that the people have "sufficient" pigs. We must first, therefore, define this quantity and discover how "sufficient" pigs are accumulated.

The Origin of the Pigs

The question of the origin of the pigs constituting the herd when it is of "sufficient size" to stage the *kaiko* is of importance here. It has already been pointed out that only 13 percent of the pigs constituting the herd at the outset of the Tsembaga *kaiko* in 1962 were purchased from people outside the local population. Accumulation of sufficient pigs to uproot the *rumbim*, therefore, seems to depend largely upon the natural increase of the local herd, rather than upon the purchase of animals from members of other local populations.

"Sufficient" Pigs

The obligations of members of the local population to make presentations of pork to others do not define the number of pigs sufficient to uproot the *rumbim* initiating the *kaiko*. Such obligations exist, and they must be fulfilled, but they set only a lower limit on the number of pigs that need be killed. This becomes clear when it is understood that except in instances in which a presentation of pork will be made in reciprocation for an equal amount of pork previously received, the size of presentations are at the discretion of the donor. It is certainly the case that he gains prestige if he makes large presentations, but there is in most cases no specific amount required to meet a man's obligations. Moreover, in those instances in which a man has specific requirements, his herd is likely to include more animals than required to meet his obligations; these will be consumed by himself and his family. The fulfillment of obligations to spirits, similarly, does not require the slaughter of a specified number of animals. Pigs must be killed in certain contexts, but the number is not specified.

It seems clear that the definition of a "sufficient" number of pigs must be sought in areas other than that of obligation to the living or dead. In earlier chapters the role of the pig in Tsembaga subsistence was discussed; it will be useful to turn here to a brief review of this information.

Perhaps the most important contribution made by pigs to the welfare of the Tsembaga is the protein they provide during periods of stress. They also provide certain other services to the human population: by eating feces and garbage they keep residential areas clean and also improve the efficiency of the use to which crops are put. The foraging of small numbers of pigs for limited periods of time not only utilizes tubers that might otherwise be wasted, but it might also benefit the arboreal component of the developing second growth.

Small numbers of pigs are easy to keep, since they are fed substandard tubers harvested along with the ration for humans. It may, in fact, be said that by eating the ration presented them the animals provide a further service to their masters: they convert tubers of only marginal usefulness as human food to high-quality protein.

There are, on the other hand, trying consequences of population expansion among the pigs. First, as the pig population expands, residential groupings become more and more fragmented. Just prior to a *kaiko* the pattern is likely to be one of subclan hamlets and scattered homesteads. The number of people with whom any individual has frequent social contacts is thus diminished. This must represent a deprivation to the Tsembaga, who, like most other people, enjoy meeting and talking, and it may also result in a lowered effectiveness of the social structure. It will be recalled, for instance, that consensus formation is achieved through informal conversations, the "talk" eventually "becoming one." Residential proximity may accelerate this process, residential dispersion inhibit it.

Perhaps more important, in a scattered settlement pattern people are vulnerable to attack. A truce prevails with the antagonists of the last fight as long as the *rumbim* remains in the ground, but there may be other groups across other borders who do not have *rumbim* in the ground and who would be free to initiate armed hostilities if a dispute should develop.

Second, the increased pig population requires additional work. The magnitude of the labor requirements of pig husbandry has already been pointed out, but some aspects of the matter must receive additional discussion.

A third disadvantage, also mentioned earlier, is that the possibilities of garden invasion by pigs are increased as the herd increases. Such events frequently result not only in damaged crops but also in slain pigs and serious fights. Slain pigs and serious fights, in turn, accelerate further the scattering of residences.

As Vayda, Leeds, and Smith have said of Melanesian pigs in general, they "apparently are in the category of those good things of which there can be too much" (1961:71). The question is not how many pigs are required to uproot the *rumbim* and stage a *kaiko;* it is, rather, how many pigs can be tolerated, and how long it takes to acquire them.

Tsembaga say that if a place is "good" it doesn't take very many years to accumulate sufficient pigs to uproot the *rumbim*. If a place is "bad," however, it takes much longer. A good place, they say, is a place where people remain well. It thus may be inferred that a good place is one in which the pig population expands because its natural increase exceeds the demands made upon it for ritual slaughter, most of which are associated with misfortune. A "bad" place, on the other hand, is one in which people often sicken or die. It is, thus, a place requiring frequent pig sacrifices. In a bad place the herd increases slowly, or not at all, or even may sometimes diminish over considerable periods of time.

While misfortune sometimes directly befalls pigs themselves, and while some pigs are killed during nonfestival periods for presentations to affines or in connection with cyclical rituals that have nothing to do with misfortune,[1] the dynamics of a Maring pig herd may be regarded as a rough index of the well-being of the human population with which it is associated.

It is difficult to translate statements concerning good and bad places into years. Informants who, it has already been pointed out, are not usually accurate in estimating lengths of time, indicate that sufficient pigs to uproot the *rumbim* may be accumulated in five or six years if a place is good.

Between two and three years is required for Tsembaga pigs to reach maximum size. Five or six years would allow both the juveniles surviving the slaughter that accompanied the planting of the *rumbim* and some of their offspring to reach maturity. A third generation might also be well grown. It may be remembered, however, that pregnancies among Tsembaga pigs are infrequent. Only fourteen litters were conceived out of approximately 100 possibilities in one year during the period of fieldwork. Infant mortality among pigs is also high. It may also be recalled that little more than 2 pigs per litter survived to the age of six months between October 1962 and December 1963. Thus, the indications are that the increase of the pig population is likely to be slow even if the ritual demands for slaughter are few. In short, a place would have to be very good indeed for a pig population of 60 juveniles, the number surviving the *kaiko* in November 1963, to expand in six years to a size approximating the 169 animals constituting the herd at the outset of the *kaiko* in June 1962.

The Tsembaga staged four previous *kaiko* during the fifty- to sixty-year period ending in 1963, which means that there were on the average approximately twelve to fifteen years between festivals. It is not clear in most cases, however, how much time elapsed between one *kaiko* and the next fight, although indications are that fighting usually did break out fairly quickly. An average of eight to

1. There are two rituals that regularly take place during the *rumbim* truce. The first of these, occurring a year or two after the planting of the *rumbim*, is concerned with transplanting *konj kump amame* (pig belly *amame*), a ritual plant, from the men's houses to the women's houses. Each land-holding group kills one pig at this time. Some time later some young men are "struck" by the shamanistic spirit, the smoke woman, and several pigs are also killed. Neither of these events, however, constitutes an important demand upon the pig herd.

twelve years between the planting of *rumbim* and the accumulation of sufficient pigs to uproot it is, perhaps, not too inaccurate an estimate for the Tsembaga.

It must be emphasized that ritual cycles of durations well outside the mean of twelve to fifteen years suggested for the Tsembaga are known. In 1963 Vayda reported that some groups had not staged a *kaiko* in over twenty years. If the number of misfortunes during a given period is either abnormally high or abnormally low, or even if there are chance variations of a sufficient magnitude in pig natality or mortality from injury or illness, the length of the ritual cycle may depart considerably from those suggested here.

PIGS, LABOR, AND WOMEN

It may be recalled that the Tomegai clan, which numbered sixteen persons, had 36.1 percent more acreage in production when the herd was at its maximum than it did when the herd was at its minimum. The difference, in square feet, was approximately 75,000, and it was estimated that this acreage required for production, harvesting, and transport a labor input of about 495,000 calories. Since the reduction in the herd amounted to eleven adult or adolescent animals, it may be assumed that the energy expense for keeping each of these pigs was approximately 45,000 calories per year, or, on an average, approximately 125 calories per day.

Since sweet potato and manioc are fed to the pigs, the burden of the labor for supplying their food falls mainly upon the women. Aside from felling the trees, making the fences, helping in the clearing of underbrush, harvesting some of the surface crops and caring for the sugar cane, men do little in the sweet potato gardens, which constitute the additional acreage. It is the women who plant and harvest the root crops and do much the greater part of the weeding and almost all of the carrying. Exact computation is difficult, but it seems fair to assume that 100 of the 125 calories that are expended as a daily average for each pig are expended by women. The question becomes, "How many pigs can a woman care for?"

In our calculation of the amount of food Tsembaga consume, it was estimated that the daily ration of a woman provides her with about 2,150 calories. Since women of Tsembaga body size need about 950 calories for basal metabolism, approximately 1,200 calories is left for expenditure in all activities. Beside caring for the pigs these include gardening for herself and her family, cooking, tending her children, and manufacturing such items as net bags, string aprons, and loin cloths, to say nothing of socializing and procreating. It is little wonder that the 66 Tsembaga females estimated to be over ten years of age were, at the outset of the *kaiko* in 1962, caring for only 169 pigs. That is, there were only 2.4 pigs of 120- to 150-pound size to each female. The range in the number of animals cared for by individual women was from 0 to 8. When the latter figure is adjusted to an equivalent in 120- to 150-pound animals, however, the range becomes 0 to 6. On the basis of the rough calculation of 100 calories per day per pig, the care of 6 pigs would demand 50 percent of the energy a woman would have available for all activity. It is not surprising that only one woman was keeping such a number unassisted, and only four women were keeping five. These numbers, it is not unreasonable to assume, approximate the maximum number that a strong adult female can care for. Since ten- to fifteen-year-old girls and old women are also included among the pig-herding females, and since these categories are not capable of as much exertion as females between fifteen and fifty, the maximum number of animals that can be cared for by the "average female" should be lowered to 4. Since there were 66 females over the age of ten among the Tsembaga during the period of fieldwork, a herd of 264 animals could have been supported. It has already been pointed out, however, that the *kaiko* started with far fewer animals, and it is probable that pig populations never approach the maximum number that can be supported by all the women in a local population. Since pigs are individually owned, some women find themselves burdened with several pigs before others have any at all. It is with the husbands of women already burdened with pigs that public agitation to uproot the *rumbim* and stage the *kaiko* apparently starts.

According to informants, as early as 1960 or 1961 certain men were already urging that a *kaiko* be staged. These men included the owners of the largest numbers of pigs. Among the reasons these men gave me for wanting to have the *kaiko* was that they were tired of planting sweet potato gardens. One, however, was frank enough to state that it was his wife who was tired, and that she had been tell-ing him so incessantly. Close observation of the domestic scene at the house of the wife of another of these men suggests that he too must have been subjected to complaints from his wife concerning the arduousness of maintaining five pigs as well as her family. Also prominent in the agitation was a widower who had been left with the equivalent of between five and six animals of 120- to 150-pound size and only a thirteen-year-old daughter to attend them. He had been forced into planting and harvesting sweet potatoes himself and on several occasions complained pathetically to me about this aspect of a widower's lot.

Men with few or no pigs responded to the talk of an approaching *kaiko* by attempting to acquire animals. Only shoats, usually under the age of three or four months, are traded, and they are always in short supply. Some were available from local litters, however, and some were obtained from other groups, particularly those residing in the Jimi Valley. As more people obtained more pigs, voices were added to those favoring the uprooting of the *rumbim.*

While it is not possible to specify a precise number of pigs that is sufficient to uproot the *rumbim,* a general statement in ecological terms can now be made. Agitation for a *kaiko* starts when the relationship of some pigs to their owners changes from one of support (emergency protein supply, conversion of substandard tubers, etc.) to one of parasitism (burdensome or even intolerable energy demands).[2] There are sufficient pigs to uproot the *rumbim* when this unfavorable change in relationship occurs in enough cases to produce a consensus within the local population.

The *kaiko* thus provides, among a group in which the slaughter of pigs is in large measure advantageously restricted by ritual to stress situations, a ritual means for disposing of a parasitic surplus of animals. In somewhat different terms it may also be said that the *kaiko* provides a means for limiting the amount of calories expended in acquiring animal protein. That the pig festivals of the Chimbu people of the New Guinea Highlands also serve to rid people of pig populations that have become parasitic has been implied by Brown and Brookfield (1959):

2. The term *parasitism* is used here, *senso latu,* to refer to a relationship between two or more individuals through which one or more are benefited or supported at the expense of others, to whom their return is significantly less than equivalent, or to whom injury is done in the process, or both.

> its [the pig ceremony's] timing depends upon the rate of growth and increase of pigs, as the main feature is a massive pig killing and distribution of cooked pork. [p. 46]

> The peak of the pig cycle . . . occurs immediately before a *bugla gende* [pig ceremony] when an adult pig population several times the size[3] of the human population makes heavy demands on land and labour, . . . much more and much stronger fencing is required to keep pigs out of cultivated land. . . . Large additional areas must be planted, not only for ceremonial foods demanded by the *bugla gende,* but also to provide more sweet potatoes for the pigs. [p. 22]

THE DESTRUCTIVENESS OF PIGS

In addition to preserving the people from further parasitism by the pigs, the *kaiko* in some instances may be a response to and a protective reaction against their destruction of gardens. This function of pig festivals has already been suggested by Vayda, Leeds, and Smith (1961):

> the pig population may . . . increase to such an extent as to become more and more a menace to the people's gardens. Among New Guinea highlanders and some other Melanesians, the fact of having a large pig population on hand is the "trigger" for holding great festivals in which so many hundreds or sometimes even thousands of pigs are slaughtered . . . whether intended as such or not by the people themselves, these massive slaughters are a way of keeping the land from being overrun by pigs. [p. 71]

Brown and Brookfield (1959:22) have emphasized the social consequences of such depredations by pigs and have suggested that among the Chimbu the disputes they cause "must be innumerable."

Because the people and their pigs were kept on one side of a steeply banked stream while all of their gardens were on the other

3. In a later publication (Brookfield and Brown, 1963:59) the same authors revise downward their estimate of the pig population at the time of the pig ceremony to approximately one pig per person. This approximates the Tsembaga figure of .83 pigs per capita.

side, there were few cases of damage to Tsembaga gardens by domestic pigs during the period of fieldwork. This geographical segregation of pigs from gardens was, however, both unusual and transitory. Informants say that there had never before been such separation, and it is unlikely that it will occur again.

Events among the neighboring Tuguma during 1963 are doubtless more representative of the extent to which the herd menaces the gardens just previous to *kaiko*. Tuguma human and pig populations were roughly comparable in size to those living on Tsembaga territory. During the period of fieldwork at least six Tuguma pigs were shot and killed after having damaged gardens. There may have been other shootings that I did not hear about, and there must have been other garden invasions in which the guilty animals were not caught.

Some of these incidents formed the basis of serious disputes. In one of these, further pig killing resulted when the owner of a slain pig took revenge by shooting an animal of the man who had killed his. One of the principals in this affair talked bitterly of moving away from Tuguma to take up permanent residence with his wife's brother, a Tsembaga. He didn't move his residence, but he did make his most extensive taro-yam garden for the year 1963 on Tsembaga land received by usufructory grant from his wife's brother, and he very clearly stated that he had chosen to plant in this location to get away from Tuguma pigs. Another man, a Tsembaga who had been living uxorilocally with the Tuguma, moved back to Tsembaga territory during 1963 after one of his pigs had been killed in someone else's taro-yam garden. It is likely that he would have eventually returned to Tsembaga anyway, but he said that it was this incident which was decisive.

Another Simbai Valley Maring group, the Kanump, were evidently experiencing tribulations similar to those of the Tuguma during 1963 just before they too expected to stage a *kaiko*. Informants there told Vayda that they wanted to have their *kaiko* soon because the pigs were ruining the gardens.

While it is the damage to the gardens that the pigs are directly guilty of, the effects of their depredations are frequently more serious than loss of foodstuffs. Garden damage sometimes leads to violence between the pig owner and the garden owner. More often it results in interpersonal taboos between the principals: they refuse to eat food cooked over the same fire, and each refuses to eat anything grown by the other. Parties to such disputes sometimes threaten to leave the territory and no doubt occasionally do. The process of residential dispersal that accompanies the expansion of the herd from the time shortly after the planting of the *rumbim* threatens to reach its logical conclusion: people may move to residences out of the territory and perhaps be permanently lost to the local group. Pigs may, in short, become competitors (for the planted crops) as well as parasites of the human population, and their competition can drive people off the land. It may be suggested that the *kaiko*, in addition to being a regulatory response on the part of the people to the parasitism of their pigs, may also be a regulatory response to the growing competitive ability of the increased pig population.

Population Densities and the Triggering of the Kaiko

Data from the Tuguma and Kanump-Kaur local populations are insufficient, but it may be that the relative importance of the two aspects of large pig herds, their parasitic requirements and their competitive abilities, vary with population densities. In less densely populated areas, or in situations like that of the Tsembaga, in which the gardens were protected from pigs, their parasitism may trigger the *kaiko*. In more densely settled areas, and in areas such as those occupied by the Tuguma and Kanump-Kaur, in which the gardens are more easily accessible to pigs, it may be their competition that determines the consensus. In other words, where population density is high and the gardens are accessible to the pigs, the number of animals required to reach an intolerable level of destructiveness might be fewer than the number required to reach an intolerable level of energy expenditure on the part of the women.

In either case the *kaiko* is likely to be triggered by population levels of pigs and people below the carrying capacity of the territory. In the case of the Tsembaga *kaiko* of 1962–63, which was clearly triggered by the parasitism of the pigs, this is obvious. It was estimated in an earlier chapter that with a human population of the size and composition displayed by the Tsembaga in 1963, the pig population could reach cyclical maxima considerably in excess of the number present without exceeding the carrying capacity of the territory. While a precise estimate cannot be made, 250 to 300 animals at the cyclical maximum would perhaps not exceed the territorial limits. A pig population within such a range would require each woman and girl to care for 4 to 4.5 animals of 120- to 150-pound size. The estimates presented here indicate that such a number is not likely to be achieved; the physiological capacity of women to care for pigs is below the capacity of the territory to provide the animals with sustenance.

Unlike situations in which the *kaiko* is triggered by the excessive labor demands of pigs, the triggering of *kaiko* by the destructiveness of the animals may be regarded as a process directly related to the density of the human population. It seems clear that with any expansion in human population there will be an increase in the number of gardens if tools, techniques, crop inventories, and planting proportions are held constant. If the area is limited, as the territories of Maring local populations are, the greater will be the number of gardens and the shorter will be the distance between them and the domiciles of pigs. That the opportunities for pig damage increase geometrically while the populations of people and pigs increase only arithmetically has already been suggested. It may be suggested here that as the human population expands, its members, increasingly troubled by the destruction of gardens, will progressively define fewer pigs, in proportion to numbers of people, as sufficient for the *kaiko*. As the human population approaches carrying capacity the number will be small. Data from local Maring populations other than the Tsembaga are insufficient, and this construction must remain hypothetical at present. It may be recalled, however, that the Tsembaga say that when, in previous times, their numbers were greater they staged the *kaiko* with fewer pigs. It may be suggested cautiously that if the carrying capacity of a territory is to be exceeded it will be exceeded by people and not pigs. The mode of population limitation posited here may not be unusual. Density-dependent processes may, and probably often do, operate at levels below carrying capacity. This has been pointed out with respect to human populations by Birdsell (1957), and Wynne-Edwards (1962, 1965) has suggested that in populations of other animals it is rare for their numbers to be affected only after increase to the level of carrying capacity. Regulation is more commonly effected at much lower population levels through density-dependent processes, such as suppression of ovulation, dispersion, and inhibition of copulation.

Whether the *kaiko* is triggered by the parasitism of the pigs or by their competition, it does seem clear that the regulation of the relationship between the pig and human populations that is accomplished by the ritual cycle helps, by periodically reducing the pig population, to keep the combined demands of people and pigs below the carrying capacity of the territory. It helps, in other words, to maintain adequate fallow periods in the secondary forests and to preserve the virgin forest cover over areas, however marginal, that might otherwise be turned to cultivation.

A further regulatory aspect of the ritual cycle may be suggested here. While the *kaiko* cannot prevent an expanding human population from exceeding the carrying capacity of its territory, it may serve to relieve local population pressure by affording increased opportunities for expansive aggression. If it is the case that, as a population grows, fewer pigs are sufficient to uproot the *rumbim*, truces should be shorter for denser populations, for the fewer the required number of pigs the less time it should take to accumulate them. To put this in the converse, the occasions when it is permissible to attack one's neighbors become more frequent with increasing population density. Data collected by Vayda should throw considerable light on this question, although unequivocal answers may never be gained due to the difficulties inherent in estimating such things as population sizes prior to fights that took place long ago and the precise intervals between fights among a people unaccustomed to reckoning in years.

While some of the suggestions that have been made here must remain hypothetical, it may be said that the ritual cycle may be regarded as a mechanism that, by responding to changes in the relationships between variables in a system, returns these variables to former and more viable levels. These variables include the number and size of the pigs and the rate of their increase, and the size, composition, and caloric intake of the human population, as well as the amount of land available to them, the distances between gardens and the domiciles of pigs, and, perhaps, other items as well. Information from the Tsembaga and other Maring groups support the suggestion made by Vayda, Leeds, and Smith (1961:72) that pig festivals "help to maintain a long-term balance between Melanesian man and the crops and fauna from which he draws his sustenance."

The Ritual Cycle of the Enemy

While the size and rate of growth of the local pig population are clearly the most important determinants of the timing of a *kaiko,* the ritual cycle of the enemy may be a perturbing factor. Both the accounts of informants concerning the time prior to contact with Europeans and the observation of events in the Simbai and Jimi Valleys in 1962 and 1963 indicate that the *kaiko* of antagonists were held at the same, or close to the same time. Rarely, it seems, were the initiation of festivals by the two principal parties to the same fight separated by more than one or two years.

Informants deny that the imminence of the enemy's *kaiko* affects their own plans, and it is likely that the near coincidence of the *kaiko* of antagonists is sometimes the result of similar processes operating in similar populations. Nevertheless, the fact that a group that has completed its *kaiko* is free to initiate hostilities, while one that has not completed its *kaiko* is not, suggests that informants' denials are expressions of ideals, rather than reports of actual motivations.

Being one year behind the enemy in initiating the *kaiko* does not expose a group to danger. Since the *kaiko* usually lasts for a little more than a year, the tardy group begins its *kaiko* shortly before or after the enemy completes its festivals. Attacks may not be mounted by a group until its *kaiko* is completed, and attacks on groups engaged in the *kaiko* seem not to have occurred. Reasons for this are not clear, but it is perhaps because *kaiko* are a matter of supralocal interest. The services rendered by any and all *kaiko* to the entire Maring and adjacent populations in terms of the movement of goods and the exchange of personnel, as well as the pork distributions, will be discussed later. It is sufficient to say here that these services are considerable and that their disruption would be regarded even by members of neutral groups as inconvenient, if not intolerable. It is doubtful that a group could rally the support of allies for an attack upon another that is engaged in a *kaiko.*

An interval of two years between the *kaiko* of enemies places the tardy group in danger, however. Their *rumbim* remains in the ground during a protracted period in which their enemies are free to attack them. The renewal of old hostilities is thus at the sole discretion of an enemy who may choose the time, place, and manner of attack. In at least one instance in Maring history an old antagonism was renewed by a group that, having finished its *kaiko*, mounted a surprise raid upon their old adversaries, whose *rumbim* had not yet been uprooted.

The implications of tardiness are apparent to the Maring. There can be little doubt, furthermore, that all groups frequently receive through neutrals information concerning events occurring on enemy territory. It may thus have been the case that in some instances groups chose to stage the *kaiko* with somewhat fewer animals than they could have tolerated.

THE KAIKO

Under the pressure of the increasing pig herd, and, possibly, in consideration also of events occurring within the enemy group, a consensus to stage the *kaiko* is finally achieved. The various events that together comprise the *kaiko* affect population dispersion, the movement of food, goods, and personnel, and both intra- and interlocal social and political relations.

Planting the Stakes at the Boundary

Preparations for the *kaiko* are initiated by planting stakes at the boundary of the territory. This ritual is performed during the earlier part of the drier season, after the trees have been cut for the new taro-yam gardens, but before planting has begun. Among the Tsembaga in 1962 this ritual took place in June or July, some three or four months before fieldwork began. Performances that were reported to be, and doubtless were, similar were observed among the Tuguma, who planted stakes for their *kaiko* in June 1963.

As in the case of all important rituals, shamans first sought the ancestors' approval for the matter at hand and asked them to designate those pigs they wanted to receive, to specify the *raku*, or pig-killing places, at which they wanted to receive them, and to appoint the day on which the killing should take place. Allies of the previous fights, in both the Tuguma and Tsembaga cases, were informed of the appointed day so that they could prepare for their parts in the proceedings.

The number of animals killed for stake planting is small. Informants agree that the Tsembaga killed only seven in 1962 and say that in earlier times only three were killed, one by each land-holding group. That a larger number were killed in 1962 was due to taboos that had come into effect during the exile of 1953–56 making it impossible for all members of any single land-holding group to share either food or cooking fires.[4]

In the addresses accompanying the slaughter of the pigs the ancestors are thanked for caring for both the people and the pigs, and they are told that there are now sufficient animals for the *kaiko* and that the people wanted to plant stakes at the boundary.

The bodies of the slain animals are cooked in earth ovens at the *raku* and apportioned to all members of the local population, regardless of age or sex. The head, heart, and lungs, however, are brought back to the *riŋgi* houses, inside of which they are cooked in *konj bint,* above-ground ovens.[5]

While the head, hearts, and lungs are cooking, the stakes are planted. The preliminary portions of the procedure need not be described here, for they are similar to those employed after an enemy has been killed during active hostilities. Spells are made upon the *rumbim, močam, deraka,* and painted stakes, and the territory is rid of both the spirits of slain enemies, who may have returned to work evil, and the corruption that such spirits spread.

4. During their exile many Tsembaga fled to the Jimi Valley, where they took refuge with the Monambant and Kauwasi local populations. These groups were enemies of long standing, and while the Tsembaga were in residence fighting between them broke out. The Tsembaga not only fought in the ranks of their hosts but, as residents, fought as principal combatants and two were killed. At the time of the Tsembaga *kaiko* in 1962–63 those who had fled to Kauwasi and those who had taken refuge in Monambant were still prohibited from eating any food grown by each other. Those who had gone elsewhere were also affected by taboos growing out of the battle between the Monambant and Kauwasi. They could eat foodstuffs grown both by those who had refuged among the Monambant and the Kauwasi but could not share cooking fires with both. If, for instance, they shared cooking fires with those who had gone to the Kauwasi they could at no time eat food cooked over the same fire as food eaten by those who had stayed with the Monambant. These taboos split all three land-holding groups, three of the five clans, and even four subclans.

5. This procedure was modified in Tsembaga in 1962. When they were returned from their exile by the Australian government the Tsembaga built no *riŋgi* houses because, informants say, the government had imposed peace and *riŋgi* could no longer be worn. The fighting stones were hung in men's houses, and it was there that the heads, hearts, and lungs were cooked.

The processions that have started from the various *riŋgi* houses and separately rid their residential areas of supernatural danger meet in the newly cut but not yet planted taro-yam gardens, which are made in one or several large clusters. During the *kaiko* year among the Tsembaga this resulted in a continuous swidden covering an area of over twenty acres. Especially careful attention is given to these gardens since it is from them that visitors to the *kaiko* will be fed. Because illness suffered by visitors would be blamed upon the locals, these gardens are deemed to be likely targets for the mischief of antagonistic spirits who sometimes make crops poisonous, and it is not considered safe to plant them until they have been thus ritually cleansed.

After they have finished with the new garden, the assemblage, led by men waving *bamp yuk* (fight packages) before them, proceeds to the enemy border. They follow the path they took to the fight ground during the hostilities, and on the way they are joined by contingents from allied groups, who also bring stakes and who, if they killed any of the enemy, have gone through similar preliminary procedures at home. On the way to the boundary all sing *welowe,* the song that was sung during hostilities on the way home from the fight ground only on the days when an enemy was killed.

At the border the new stakes are planted. These, together with the old stakes and growing *rumbim* to which the cleansing objects are tied, form a gate to the territory. Spells are made to send both the enemy spirits and their corruption back to the enemy territory from which they came. Large trees are then designated for felling, one for each enemy killed. The groups responsible for the killings, both local patrilineal groups and allies, chop them so that they fall across, or at least in the direction of, the border. While they chop, the men sing *welowe,* and the fight magic men rub the trees with their fight packages.

Some informants say that the act of cutting down the trees is simply a celebration of the power of the fight packages, with the aid of which men "as big as trees" were killed. Others say that the trees are being offered to the spirits of the slain men to use as residences. Cutting them so that they fall across the border, moreover, sends the spirits (*rawa*) of the trees[6] across the border, where, it is

6. This is the only context, to my knowledge, in which *rawa* are attributed to trees.

hoped, their presence will induce the spirits of the slain men to remain. The procession then returns home, the allies dispersing to their various territories, the locals to their several *riŋgi* houses.

With the opening of the ovens at the local *riŋgi* houses the taboo against trapping marsupials is abrogated.[7] *Andik meŋ*, "turning word," spells are made on cuttings of the shrubs *gañiŋgai* (*Elatostema* sp.). These are held in the steam of the newly opened oven by a fight magic man who, in a conversational tone, addresses both the living and the deceased and recounts the story of the fight, the subsequent planting of the *rumbim*, and enumerates the taboos that have remained in effect ever since. He says that now that the stakes have been planted the men would like to trap marsupials so that drum heads can be made, but when they fought, he continues, they said, in obedience to tradition (*nomane*), that they could not trap marsupials and these words remain inside of them. Before they set their traps, therefore, they must rid themselves of these words. The *gañiŋgai* is then passed among the men and boys, each throwing back his head and brushing himself upward from the navel to the mouth. While brushing, each announces that he is ridding himself of the words of the taboo and he makes a spitting sound. The pig is then eaten and the ritual concluded, therby terminating the "closed season" on marsupials.

There are two other aspects of stake planting that should be commented upon. First, the assemblage that joins to plant stakes is supralocal. The territory being redefined by the planting of the stakes belongs to only one of the assembled groups, but the participation of several of them seems to signify joint defense of the territory. Stake-planting rituals may perhaps be regarded as cyclic ratifications of mutual assistance agreements between members of several local populations.

They also may be regarded as display behavior. Every participant in the stake-planting procession has an opportunity to gauge the size or strength of the entire assemblage and of its constituent units, and the enemy is also exposed to this display. Enemies are said to be afraid to come to the border, or anywhere near it, to witness the spectacle. They therefore view the procession only from a distance at best, and if the terrain is very broken, as at the Tsembaga-Kun-

7. I did not witness this part of the ritual in Tuguma in 1963. I have, however, seen similar performances in other contexts on several occasions.

dagai border, or thickly wooded, as is the border between the Kanump-Kaur and the Tuguma, the enemy does not see the procession at all. He hears it, however. Some enemy men, while remaining out of eyeshot, come close to the border to make counter-magic against the evil that is being sent to them by the stake planters. Even those who remain at a greater distance hear the procession, for the sound of two hundred men or more singing and shouting battle cries carries far in the quiet valley.

The enemy also hears about the procession either from eye witnesses or, more likely, from those to whom eye witnesses have spoken. These reports are inevitably impressive simply because the Maring have no terms for quantities larger than twenty. Most second- or third-hand accounts of events among the Maring are exaggerated, and it is likely, despite the lack of numerical terminology, that these are too. At the least, the enemy gets the impression that a very large number of men participated in the ritual, and this might serve to temper any bellicose plans he might entertain for the future.

The most important aspect of the ritual is not that of agonistic display, however; it concerns where the stakes are planted. If the enemy was not driven off its territory in the last fight, but remained to plant *rumbim,* or if, having been driven out, the enemy has returned and planted *rumbim,* the stakes are planted at the boundary that existed before the fight.

If, however, the enemy was driven out of its territory and never returned to plant *rumbim,* the procession does not stop at the old border. It proceeds into the territory of the former enemies and the stakes are planted at a new location. A new boundary is thus established, incorporating into the territory land previously held by the enemy.

The Tsembaga and most other Marings say that fights do not take place over land and that land occupied by other groups cannot be annexed. To signify its occupation, however, a group must plant *rumbim* upon its land. Areas annexed in stake-planting rituals are areas upon which no *rumbim* is planted; they are, therefore, not lands belonging to the enemy, but lands the enemy has presumably abandoned—they are vacant.

A simple rule may be presented here: *if one of a pair of antagonistic groups can plant its stakes before its opponent can plant its* rumbim, *it may annex land previously held by its opponent.*

It is not only the vanquished who have abandoned their territory; it is assumed that it has now been abandoned by the ancestors of the vanquished as well. The surviving members of the erstwhile enemy group have by this time resided with other groups for a number of years, and all or most of them will already have had occasion to sacrifice pigs to their ancestors at their new residences. When they do so they invite these spirits to come to the new place of the living, where they will continue to receive sacrifices in the future. Ancestors of vanquished groups thus relinquish their guardianship over the territory, making it available to the victorious groups. Meanwhile, the de facto membership in the groups with which they have taken refuge is converted to de jure membership. Sooner or later the host groups will have occasion to plant *yu miñ rumbim,* and the refugee men as coresidents will participate, thus ritually validating their connection to the new territory and the new group. A second rule of population redistribution may thus be stated: *a man becomes a member of a territorial group by participating with it in the planting of* rumbim.

There are two processes that modify or complicate the lasting effects of these rules. The first is the partial reoccupation of its territory by a routed group before the victorious group plants its stakes. It is often the case that some members of a routed group will take refuge with an adjacent group, under whose cover they will reoccupy some of their lost territory. This has happened at least twice in Tuguma-Tsembaga history. When the Dimbagai-Yimyagai, many years ago, were first driven from their land by the Tsembaga on their west, some took refuge with the Dinagai, then a local population immediately to their east. As members of the Dinagai they reoccupied the eastern portion of their territory. The Merkai clan of the Tsembaga recognized this as an annexation of part of the territory by the Dinagai, with whom they were friendly, rather than as reoccupation by their former foes. They therefore planted their stakes to incorporate only part of the territory from which they had driven their enemies. Similar processes took place following the rout of the Tsevent by the Tuguma.

Second, while the planting of stakes to annex abandoned territory provides what might be characterized as de jure rights to that territory, anxieties seem to remain with the conquerors concerning the use of this land for gardens and residences. These anxieties apparently become explicit during times of misfortune. During 1962–63 almost all of the Tsembaga residences were on land that had, at one time, belonged to the Dimbagai-Yimyagai. In 1962 five Tsembaga men and one Tsembaga woman of middle age or younger died of illness, and in early 1963 two more young men, a young woman and a child, became sick and quickly died. After each death there was much talk of giving the land back to the Dimbagai-Yimyagai. Some people said that the Dimbagai-Yimyagai ancestors, wanting to return to their own land, were sending illness. Others blamed the deaths on living Dimbagai-Yimyagai who, they said, were sending the illness magically. Consensus in favor of abandonment seemed to form rapidly, but it could not be acted upon immediately because the *kaiko* dance ground, in addition to the residences, happened to be on the old Dimbagai-Yimyagai territory. Their departure, therefore, had to be delayed until after the *kaiko*.

This delay permitted the consensus to dissolve. After most of the deaths, talk in favor of removal seemed to remain firm for one or two weeks. After a month or a little longer, however, those who had advocated removal were frequently surprised when they were reminded of their position. They would point out that the Dimbagai-Yimyagai had all planted *rumbim* elsewhere, that the Dimbagai-Yimyagai ancestors had long ago vacated the territory, and that the matter was ancient history. Statements to the effect of, "It was our fathers who drove them out, and now we ourselves have children," were common.

If they hadn't been detained by the presence of the dance ground, however, the Tsembaga may well have given the territory back to the Dimbagai-Yimyagai. Because of Australian pacification they no longer needed to fear the presence on their border of a reconstituted Dimbagai-Yimyagai, who at any rate would not be numerically strong. Furthermore, all indications are that the Tsembaga had experienced considerable depopulation since they annexed Dimbagai-Yimyagai land. They now had, or at least said they had, sufficient land without the old Dimbagai-Yimyagai territory, and estimations of carrying capacity support their view.

History thus indicates that annexation of land through conquest and the subsequent stake planting is reversible through peaceful processes. It further suggests that the mechanism through which the annexation is reversed, involving both native theories of disease etiology and native behavior when suffering an especially high incidence of death, depends on the population density. Groups which, because of depopulation, no longer need land they have conquered may abandon it under the impact of further depopulation.

Preparing the Dance Ground

In the weeks before and just after the planting of stakes, the settlement pattern changes from one of scattered homesteads and subclan and clan hamlets to one of relative nucleation around a traditional dance ground. This, of course, results in the concentration of a large number of pigs in a relatively small area. This concentration is temporary, usually lasting little more than a year, and its ill effects upon second growth are therefore limited. It may also be that the threat of pigs to the gardens during this period is offset by the practice, during *kaiko* years, of planting the taro-yam gardens in large clusters. Clustered gardens require fewer linear feet of fence per unit of cultivated area because of a reduction in total periphery. Fencing around the taro-yam gardens in *kaiko* years seems to be exceptionally stout, and it may be that the reduction in linear requirements for fencing permits this sturdier construction.

It may also be usual for gardens during festival years to be planted at a greater distance from the residences than in other years, thus placing them beyond the usual daily range of pigs. Almost all of the gardens planted by the Tsembaga in 1962 were at least a thirty minutes' walk away from the residences. It is not possible, however, to generalize from Tsembaga procedures in 1962, because, as has already been mentioned, their settlement had been, contrary to usual arrangements, nucleated for some years before the *kaiko,* and most of the arable land close to the settlement was under secondary forest too young to cut. The Tuguma gardens planted in 1963, the year of the Tuguma *kaiko,* were generally not as far from the settlement as those of the Tsembaga.

While the Tsembaga settlement was nucleated long before the *kaiko,* it was far from any of the old dance grounds. Instead of moving their residences to the proximity of one of the traditional dance grounds they prepared a new one within the area encompassed by their settlement. The work was considerable, for the site sloped at an angle of more than twenty degrees. Instead of the easy task of removing weeds and saplings from a place previously used for dancing, they were faced with the necessity of leveling a considerable area. Since no one could do more than exhort others to work, and since enthusiasm for the task waxed and waned, it took months to level (roughly) an area of about 150′x200′. Improvements were being made and additional areas were being cleared on the slope deep into the *kaiko*.

The work was not confined to preparing the ground. Two large houses, each about 25′x35′, and 8′ or 9′ high at the center, also had to be built at the edge of the dance ground to accommodate visitors if it rained. Two houses, rather than one, are necessary at almost all *kaiko* because a man may never enter a house into which an enemy has ever set foot, and some of the groups to be entertained are almost always enemies of others.

Immediately after the stake planting and before leveling had actually begun, a fence was built around those portions of the area where the ground did not fall off so steeply as to make approach difficult. At the gate, the root of a variety of *rumbim* called *dawa*, the leaves of which are used to cover the buttocks when dancing, was buried after a spell was said over it. This ritual, which included an address to the ancestors, had two purposes, according to informants: to induce the ancestors to look kindly upon the dance ground, that both local and visitors alike might dance strongly and that their drums might sound rich; and to keep within the fence the *miñ*, the life stuff, of the local girls while the *kaiko* continued. The men often express the fear that some of their unmarried girls might elope with visitors whose strong dancing and finery had captured their fancies.

Uprooting the Rumbim

With the ripening of the first *marita pandanus*, a yellow-fruited variety called *yambai*, preparations for uprooting the *rumbim* begin. In Tsembaga territory this fruit is ready to eat in late August or early September, by which time two months or a little more has elapsed since the stake planting, and the bulk of men's work in the new gardens should be finished or close to finished.

When the *yambai,* which is a scarce variety, becomes ripe it is eaten once, after which a taboo on further consumption of it is assumed by everyone. Even the word for *marita, komba,* must be avoided, and a circumlocution is used until the taboo is renounced during the rituals immediately preceding the uprooting of the *rumbim. Marita,* informants say, is associated with the spirits of the low ground, and it is now time to trap *ma* (most marsupials and perhaps some giant rats are included in this category), which, it may be remembered, are said to be the "pigs of the red spirits." Undivided attention must now be turned to these spirits, and the ingestion of *marita,* or even the enunciation of its name, would make this difficult or impossible, either because the act would annoy the red spirits or because it would have some direct effect upon the body or mind. This temporary renunciation of *marita,* it must be said, costs the people little, for after the few *komba yambai* fruits are consumed virtually no *marita* becomes ripe until middle or late October.

Ma trapping is carried on separately by the least inclusive agnatic units, clans in some cases, subclans in others, in their own *komoŋ,* tracts in the high-altitude virgin forest which are said to be the homes of their red spirits. While at ordinary times a man may set his traps anywhere on Tsembaga territory, he may now set them only in the *komoŋ* of his own minimal agnatic unit because it is only with one's own ancestors that "pigs" may be exchanged. Before the traps are placed, shamans contact the red spirits, saying to them that the pigs of the living are now sufficient in number to be exchanged for the red spirits' pigs. They ask the smoke woman to designate those trees in which the red spirits will place the marsupials they wish to give to the living, and it is in these trees that the traps are placed. During the course of the period the smoke woman is asked from time to time to designate further trees.

Adolescent boys and men through middle age participate in the trapping. Because this is an activity associated with the red spirits, the trappers are subject to many of the same taboos as they are during warfare. They may not have sexual intercourse, or even touch women, nor may they eat food prepared by women. They should not set foot in the *wora,* the lower portion of the territory, and food grown in the *wora* should be avoided. They may not leave the territory, and they may not share food cooked over the fires of other local populations.

The trapped *ma* are skinned, and then smoked by the old men and the boys too young to join in the trapping. Some of the animals taken early are preserved for two months or more. Informants say that it is only in the context of this ritual trapping period that meat is thus preserved; smoking was never observed during the fieldwork period.

Special smokehouses are built for this purpose. While the actual trapping is conducted independently by minimal agnatic units, smokehouses are usually built, informants say, by sub-territorial groups. There was some modification of this among the Tsembaga in 1962, however, because of the food and fire taboos, mentioned earlier, which split these groups.

When the variety of *marita* called "*peŋgup*" ripens in mid-October the men stop their trapping and prepare to uproot the *rumbim.* The preparations are elaborate, for pigs must be slaughtered at the *raku*, which means that shelters and above-ground ovens must be built, and vegetables and firewood must be obtained. Allies also must be informed so that they may attend.

The rituals performed at the *raku* on the day before the *rumbim* is uprooted are both elaborate and exotic. Pigs are sacrified for both the spirits of the high ground—the red spirits and the smoke woman —and the spirits of the low ground. In the address preceding the sacrifice for the spirits of the high ground, in addition to being thanked for their help in warfare, the spirits are thanked for the *ma* they have provided and are told that they will now be given pigs in exchange for them. Among some Maring local populations cassowaries are also commonly sacrificed to the red spirits, but the Tsembaga keep very few of these birds.

After the pigs are killed, oven stones are laid on a large fire to heat while the butchering of the pigs and the preparation of the smoked *ma* for cooking proceeds. When these preparations have been completed two large red *pandanus* fruits, harvested with great care on the previous day from groves in the low ground, are brought forth. The groves from which the fruits are obtained may either be those planted by persons now deceased, or those in which the remains of dead are buried. A procession forms, composed of all those present.

Led by two men who continually raise and lower the *pandanus* fruits they are carrying, the group circles the fire chanting:

komba ku komba *yaŋga yaŋga muŋga muŋga*
kam ku komba *yaŋga yaŋga muŋga muŋga*
bri komba *yaŋga yaŋga muŋga muŋga . . .*

(*Marita,* ascend to and come down from *komba ku*
Marita, ascend to and come down from *kam ku*
Marita, ascend to and come down from *bri . . .*)

Komba ku and *kam ku* are high places on Tsembaga territory where the smoke woman is said to dwell. *Bri,* another of her dwelling places, is a mountain in the Jimi Valley. The chant continues, naming many other high places in the Jimi Valley that are said to be homes of the smoke woman, and becomes more frenzied as it proceeds. The voices of some of the men break into sobs. When the catalog of the homes of the smoke woman is almost completed the procession halts. All males take hold of one of the *marita* fruits, while the females grasp the other. When the chant is completed, one of the adult men seizes the *marita* of the males and with it leaps onto the oven stones that have now been heating for well over one hour. Bounding up and down on the hot stones, he stabs the fruit with a cassowary bone, then leaps off. He repeats the performance with the *marita* of the females.

The various foods are now put into the ovens. The pig dedicated to the red spirits and the smoke woman is cooked in an oven constructed above ground, while that for the spirits of the low ground is placed in an earth oven. Marita and marsupials are cooked together, but two special ovens are made for this mixture because fight magic men suffer an enduring taboo on consuming some marsupials with women.

When the ovens are opened, the men who have participated in warfare gather around the above-ground oven. The head of pig cooked in it is raised high while one of the men recounts the story of the last fight, thanks the spirits of the high ground for their assistance, tells them they are being given this pig and that more will be given to them at the end of the *kaiko.* The address concludes with the request to the spirits to take this pig, which is now being offered them, and return to their high places. The oven of the spirits of the low ground, that made in the earth, is opened without ritual.

All of the greens and some of the flesh from both ovens is consumed at the *raku,* but most of the flesh is brought to the residences for later consumption or, in the case of the pig cooked in the earth ovens, for distribution to members of other groups.

Several taboos must be abrogated before the *marita* and *ma* may be eaten. One of these applies to novice shamans, who have been prohibited from eating any *marita* since being "struck" by the smoke woman, in some cases years before. The others apply to all persons. These are, first, the prohibition, in effect for several months, against the consumption of *marita* during the trapping period. The other, in effect since the last fight, has forbidden the cooking and consumption of *marita* and *ma* together. All three of these taboos are ended by spitting out a mouthful of *pandanus* seeds and throwing away the tails of the *ma* in the nearby forest. This latter act is also said to ensure the future proliferation of the *ma.*

After the taboos have been nullified some *pandanus* oil is rubbed on the legs and buttocks of all persons so that their legs will be strong, and on the bellies of the females so that they may be fertile. The cassowary bone previously used to pierce the fruit is used as a spoon to feed each person his first mouthful of the *marita.* The consumption of the *marita* and *ma* is the last ritual act of the day.

The day-long proceedings at the *raku* have two stated purposes. On the one hand they enhance fertility, and on the other hand they nullify a number of taboos. In addition to those restricting the consumption of *marita* and *ma* by all or some categories of persons, the taboos renounced include those, in effect since the last fight, that have prohibited all men who have worn *riŋgi* from sharing with women certain foods including certain species of *ma* and certain varieties of *marita* as well as sugar cane, bananas, and *pitpit* having red skins. The men who wore *riŋgi* also renounce their taboos against the consumption of certain other foods, notably soft varieties of *Dioscorea* grown in the *wora,* the lower part of the territory. The taboos on beating drums and eating pigs slaughtered in connection with the *kaiko* of other local populations are also terminated.

It should be mentioned that these taboos (except that against drum beating) are not renounced by all men. Fight magic men are burdened with them for life. Furthermore, the taboos against the consumption of snakes, lizards, frogs, and other "cold" animals also remain in effect for all men for life.

It was suggested earlier, in reference to their assumption, that some of these taboos serve to segregate ideologically those components, both living and dead, of the total community that are associated with warfare from those associated with peaceful activities. This segregation, it was suggested, is an aspect of the local population's debtor relationship with both the ancestors and allies from whom they received assistance in warfare. The termination of these taboos and, thus, the partial reintegration of the previously segregated elements is concomitant with the partial repayment of the outstanding debts.

Explanations of the significance of much of the ritual could not be elicited from the Tsembaga, but the forms of some of the procedures, particularly those concerned with *marita,* support the notion of reintegration. *Marita* is clearly associated with the spirits of the low ground. Indeed, the fruits used in the ritual must be taken either from trees planted by the deceased on low ground or from groves where the remains of the deceased lie. But before it is placed in the ovens the *marita* is offered to the smoke woman, a spirit of the high ground in whose name novice shamans had suffered an absolute prohibition against its consumption. The fruit, moreover, is pierced with a cassowary bone before cooking. Cassowaries are associated with the red spirits, for whom *norum-kombri* (orchid sp.—cassowary), is often used as a term of address. *Ma,* now cooked with *marita* for the first time since the last fight, are also associated with the spirits of the high ground, being regarded as their pigs.

The relationship of the living to the two categories of spirits seems to undergo transformation in the new reintegration. This is signaled by the ritual consumption of the *marita,* an act, the Tsembaga say, which is like taking the spirits of the low ground inside of one. This is not said of the consumption of the *ma.* These animals are regarded to be the receipts of an exchange with the red spirits, and this is made explicit in the address preceding the slaughter of the pig dedicated to them. What may be regarded as an analogous communion with the red spirits took place when, before the ax fight, the fighting men put on *riŋgi* and took these spirits into their heads. Performances undertaken now, renouncing taboos that were assumed with the *riŋgi*, may be regarded as expunging some of the vestiges of that communion. This interpretation is supported by the address to these spirits when the head of the cooked pig is offered to them: they are asked to take the pig being offered them and to leave. This behavior in reference to the spirits of the high places, in short, seems to suggest not a communion, but its opposite: the expulsion of spirits with whom a burdensome communion was previously effected. It further supports the interpretation that the abrogation of the taboos associated with the wearing of the *riŋgi* requires the transformation of the simple debtor-creditor relationship that has prevailed since the last fight into a relationship that, although obligations remain, is becoming more equal. That the ability to transform the relationship from one of debt to one of equivalence depends upon the demographic and ecological success of the local population is apparent in light of earlier discussion.

The number of pigs killed at this time was substantial—the Tsembaga killed thirty on November 1, 1962. Those killed for the spirits of the high ground, in most cases smaller animals, were consumed by the local population. With the exception of the heads and entrails, pork derived from those killed for the spirits of the low ground was carried back to the residences, where it was presented to the waiting allies. For the most part this flesh was given unceremoniously by each man to those men among the allies, usually his own affines or non-agnatic cognates, who assisted in the fight because of their ties to him. The belly fat was first removed, however, to be salted for formal presentation.

On the following day, after the salted belly fat was presented to the allies, the *rumbim* was uprooted.[8] Both allies and local men were in attendance. *Andik meŋ* spells were first made upon the stacked drums, so that they might be beaten and so that they might sound well, after which they were returned to their owners. Painted stakes, about six inches long, upon which spells had also been made, were then buried near the gate of the enclosure. Now that the *rumbim* was to be uprooted these would protect the *miñ*, the life stuff of the

8. This ritual was performed only seven days after our arrival among the Tsembaga and I was permitted to watch only from a distance. It was also not possible for me to attend its performance at Tuguma; therefore this account depends in part upon informants' statements.

men, preventing both the spirits of enemies and the corruption that flows from them from entering the house enclosure.

After the stakes were buried, the *amame* was unceremoniously pulled out. The *rumbim* was then uprooted by the man who planted it. This task always, it is said, falls to the man who planted it if he is alive, and to his son if he is not. A digging stick upon which *andik meŋ* spells had been made was used. The address, broken by sobs, recounted the history of the planting and the fight preceding it. Both categories of ancestors were told that they had been given some pigs, but only a few; later, at the end of the *kaiko,* they would be given more. In previous times, informants say, the red spirits were also told that after the *kaiko* their deaths would be avenged. Since pacification, apologies are made to the red spirits because revenge is no longer possible.

While the *rumbim* was being uprooted, bamboo was heated over a fire. When it exploded the *rumbim* was torn from the ground and all the men pounded their drums, yelled, and stamped their feet. Led by men carrying the *rumbim,* they all charged over the stile and headed for the territorial boundary. On the way they were joined by groups who had uprooted *rumbim* at other *riŋgi* houses, as well as by a contingent from Tuguma bringing bespelled stakes.

While most of the array proceeded to the boundary where the stakes were planted, some fight magic men dropped out to dispose of the *rumbim* and *amame* elsewhere. These are traditionally taken to shallow places in certain streams where they are placed on flat rocks with their roots covered by water but with their foliage on dry land. They are oriented so that the foliage points in the direction of the enemy, while the roots point in the direction of the local settlement. Spells are made upon them so that the desiccation which, it is said, will overtake the foliage, will go to the enemy, bringing him illness and death. The roots, on the other hand, rot, and from rot comes new life. Since the roots are pointed in their direction it is the local population that benefits in fertility and growth from their decay.

After the stakes had been planted at the boundary and the *rumbim* and *amame* disposed of in streams, the procession returned to the dance ground. Dancing, accompanied by drumming and singing, continued throughout the night. The *kaiko* had begun.

Kaiko wobar and kaiko de

The year-long *kaiko* is divided into two periods, the first called *wobar*, the second *de*. It may be recalled that these are the names of the songs that were sung on the way to the fight ground, *wobar* during the earlier "nothing fight" and *de* for the ax fight. During the *wobar* period of the *kaiko* only *wobar* may be sung; during the *de* period both *de* and *wobar* may be sung.[9] The two stages of the *kaiko*, informants say, recapitulate the two stages of warfare. The earlier, less important, *wobar* phase continues while work remains to be done on the dance ground, and, more importantly, until there are enough large taros and yams in the new gardens to entertain visitors properly. When there are enough tubers fit for presentation, the rituals inaugurating *kaiko de* may be performed. The first taro and yams may be harvested about six months after planting, but at least another month is required for them to become sufficient in size and number. The Tsembaga had enough of these tubers in late March or early April 1963 to perform the *de* rituals, but were delayed by a number of extraneous events until early May.

As the two stages of the *kaiko* recapitulate the two stages of warfare, so do the rituals preceding the two stages of the *kaiko* reverse the effects of the rituals undertaken in connection with the two stages of fighting. In the case of the bow and arrow fight the preliminary rituals, which are minimal, are reversed, according to most informants, as one of the effects of the uprooting of the *rumbim*. Two informants also stated that as part of the reversal procedure the arrows and spears used in the small fight are brought to the *raku* on the day before the *rumbim* is uprooted, where they are treated with the ritual *marita*. I did not witness such a performance, however, among either the Tsembaga or the Tuguma.

The details of the rituals inaugurating *kaiko de* need be sketched in only briefly. As always, the shamans contact the smoke woman to ask the spirits' approval and to ask her to designate the *raku* at which the pigs should be killed.

9. *De* and *wobar* refer to melodies, for each of which there are standard refrains but innumerable verses, with new ones constantly being composed or imported. Not all of these are in the Maring language; many of those sung to the *wobar* melody are in Karam, while some of the *de* verses are in Narak. The *de* refrain, moreover, is reported word for word by Luzbetak (1954) in his description of a pig festival among the Nondugl people of the Eastern New Guinea Highlands.

Five animals were slaughtered by the Tsembaga, all of which were consumed by themselves. During the night the *rumbim* that was planted outside the men's house on the night before the ax fight is uprooted. Extraction and disposal spells are then made not only on the usual stakes, *rumbim* brooms, wild taro leaves, and vine whips, but also on bamboo torches, which are then lit on fires made inside the *riŋgi* houses. The dance ground and the entire residential area are then rid of the spirits of slain enemies and the corruption emanating from them as well. When the men return from depositing the uprooted *rumbim* and other objects at the border, the fighting stones are removed from the center post of the *riŋgi* house or, in the case of the Tsembaga, from the rafters of men's houses, and placed on the low tables from which they were taken years before during the night before the ax fight.[10]

The lowering of the fighting stones is not accompanied by the immediate termination of remaining taboos. It does not now become possible, for instance, to enter the territory of the enemy or to speak to the enemy, much less make a permanent peace. However, lowering of the stones is a prerequisite to making peace at a future time and also to the more immediate trapping of eels, which precedes the final pig slaughter. Eels are said to be the pigs of the spirits of the low ground, for which pigs of the living will be exchanged. As it was necessary when *ma* were being trapped to avoid contact with the spirits of the *rawa mai*, so is it necessary before taking eels to continue the process of turning away from the red spirits by lowering the stones. It was mentioned previously that hanging the stones signified the assumption of a debt to both allies and ancestors. Their lowering, conversely, signifies that the debt is soon to be paid.

On the following day another ritual is usually performed. The allies assemble on the dance ground, and small trees that were, when the area was cleared, allowed to remain growing for his purpose are assigned to each group responsible for slaying an enemy in the last fight. After these trees are rubbed with the fight packages they are uprooted while the men sing *welowe*, the killing song.

The killers, or, if they are no longer alive, their sons, are then carried around the dance ground on the uprooted trees, and *de* is

10. I was at the Simbai patrol post recuperating from illness when *kaiko de* was inaugurated by the Tsembaga. This brief sketch is based upon informants' accounts and the account of my wife.

sung. All informants agree that this is a celebration of the power of the fight package, and some say that the spirits of the enemy slain actually inhabit the trees, which apparently symbolize them. The rite also, of course, honors those men who have killed enemies in the service of the local population.

The Tsembaga chose in 1963 to forego this ritual. Men said that they would be ashamed to uproot trees after a fight in which they had killed only two of the enemy while losing twenty themselves.

Differences between the style of entertainment of visitors during the *wobar* and *de* periods are actually minor. During the latter period there is a second melody at the disposal of the dancers and the food presentations become more elaborate as taro and yams become increasingly available. This distinction becomes blurred, however, if the inauguration of *de* is delayed. The Tsembaga were including taro and yams in their presentations for more than two months before they undertook the *de* rituals.

A Kaiko Entertainment

The occasions on which friendly groups are entertained,in addition to the year-long festival, are called "*kaiko*," a term also used to refer to the dancing, which is one of the features of these events. During the course of the year the Tsembaga entertained thirteen other local groups on fifteen separate occasions, exclusive of the final *konj kaiko*, "pig *kaiko*," which terminated the entire festival. On several of these occasions two or more local groups were entertained at the same time, and three local groups attended more than once.

INVITATIONS AND PREPARATIONS

The atomistic nature of the organization of local populations is clearly expressed in the extension of invitations to a *kaiko*. Invitations are not extended by a local population as a whole to another local population as a whole, but rather by one or several individuals in one local population to one or several individuals, either kinsmen or trading partners, in another. Usually the men who have such connections in a particular group will together decide when to issue invitations to these kinsmen or trading partners. They may, in their planning, take into consideration commitments made previously by other members of their local population, but decisions are likely to be made without the counsel of others. Indeed, in instances in which kin or other ties with another group are few, the issue of an invitation may be the decision of a single man who has taken no counsel but his own.

Although invitations are formally extended by individuals to individuals, they are in fact invitations to entire local populations, for the invited are expected to bring at least some of their coresidents. Indeed, the formally invited men would be ashamed to attend if they could not bring with them dancers in a number commensurate with the size of their local population and with the strength of the relations between their local population and that of their hosts. The invited men gain prestige by making a strong showing, and it is to this end that they themselves often extend invitations to members of yet other local populations, to "help them dance" at the *kaiko* to which they have been invited.

As the extension of invitations to a *kaiko* reflects the atomistic character of local organization, so do preparations for these events. It is the responsibility of those men who have kinsmen and trading partners among the visitors to accumulate the firewood and sugar cane required for their entertainment, and of their wives to harvest and prepare the tubers and green vegetables with which the visitors will be fed. Others may contribute effort or garden produce to the preparations, but such contributions are phrased as assistance to those upon whom the responsibility for entertainment falls and through whom the food presentations are made. While some of those who have no direct ties to the visiting group may assist those who do, it is usually the case that some people take no part whatever in the preparations.

THE ENTERTAINMENT

Although days are appointed, temporal arrangements remain vague. Young men are sent in advance to keep the hosts informed of delays and revised times of arrival of the rest of the group. On the day on which the hosts are reasonably certain that the visitors will arrive, extraction and disposal magic is performed on the dance ground. This is usually a perfunctory proceeding, and after the spells are made upon the usual objects the task is turned over to whatever young men happen to be in the vicinity. The object of the extraction and disposal magic is to rid the dance ground of both *tukump*, the corruption that flows from spirits, and *kum,* which is similar but is introduced into the ground by living sorcerers.

After the dance ground has been treated, those men who wish to participate in the dance bathe in streams and then return to the men's houses to adorn themselves. Those who ready themselves for the dance include all or most of the young men and adolescents, and at least those older men who have direct kinship or trading connections with the visiting group. Other older men without direct connections to the visitors may also participate, either to assist those who do have direct connections or simply because they wish to dance.

As invited groups receive support or assistance from other groups when they go to a *kaiko,* so does the host group. On most occasions at Tsembaga young men of Tuguma would swell the ranks of the Tsembaga dancing contingent.

Adornment is painstaking, and men often take hours to complete their dressing. Pigments, formerly earth colors of native manufacture, more recently powders of European origin, are applied to the face in designs that are subject to frequent changes in fashion. Beads and shells are worn as necklaces, and garters of small cowries encircle the calves. The best orchid fiber waistbands and dress loin cloths enriched with marsupial fur and embellished with dyed purple stripes are put on. The buttocks are covered with masses of accordion-folded leaves of a *rumbim* called "*kamp*" and other ornamentals. A bustle, made of dried leaves obtained in trade, which rustles during dancing, is attached on top of the mass of *kamp* leaves.

Most attention is given to the headdress. A crown of feathers, eagle and parrot being most common, encircles the head. The feathers are attached to a basketry base, which is often hidden by marsupial fur bands, bands made of yellow orchid stems and green beetles, or festoons of small cowrie shells. From the center of the head rises a flexible reed, two or even three feet long, to which is attached a plume made either from feathers or an entire stuffed bird. The lesser bird of paradise, the Princess Stephanie bird of paradise, the greater sickle bill, and (?) Pescot's parrot are most common. Plumes, particularly those of the King of Saxony bird of paradise and the racket-tailed kingfisher, are worn through the pierced septum from which also may be suspended a shell disk and gold-lip crescent. Unmarried girls, some as young as nine or ten years of age, also may be adorned.

When their preparations are completed the dancers, both men and girls, congregate at the dance ground, where they dance for practice or pleasure before the visitors arrive and where they may be admired by spectators who are already assembling. These include the local married women and men from many other places who, having heard of the *kaiko,* have come to watch and to trade.

The visiting dancers signal their approach by beginning to sing. When they reach a point about one hundred yards from the gate, the local dancers retire to a vantage point just above the dance ground, where their view of the visitors is unimpeded and where they continue singing. The visitors approach the gate silently, led by men carrying fight packages, swinging their axes as they run back and forth in front of their procession in the peculiar crouched fighting prance. Just before they reach the gate they are met by one or two of those locals who have invited them and who now escort them over the gate. Visiting women and children follow behind the dancers and join the other spectators on the sidelines. There is much embracing as the local women and children greet visiting kinfolk. The dancing procession charges to the center of the dance ground shouting the long, low battle cry and stamping their feet, magically treated before their arrival both to counteract any *kum* or *tukump* which may linger in the ground and to enable them to dance strongly. After they charge back and forth across the dance ground several times, repeating the stamping in several locations while the crowd cheers in admiration of their numbers, their style, and the richness of their finery, they begin to sing. Their first song, to the appropriate melody, either *de* or *wobar,* should be one composed in honor of the occasion.

While the visitors are thus displaying themselves, fight magic men among the hosts apply magic to the feet of the local dancing contingent so that they may dance strongly, and fight packages are applied to their headdresses, so that the beauty of the feathers may attract the fancies of the visiting girls. Bamboo is heated while the presiding fight magic man, his voice broken by sobs, asks that both categories of spirits help the locals, that their dancing not be outdone by the visitors'. They also ask that visiting girls be attracted by the dancing and the feathers of the locals, but that the local girls remain unmoved by the charms of the visitors. When the bamboo explodes the locals charge onto the dance ground pounding their drums and singing. Their numbers include at first the girls who have decorated themselves for the occasion. They dance inside the male ranks for a few minutes before retiring to the sidelines. The formations of the locals and the visitors remain separated at first.

It is usually late afternoon when the visitors arrive. Just before dusk the locals stop dancing and assemble the food that has been prepared for the visitors in the middle of the dance ground. It includes bundles of sugar cane, net bags full of cooked tubers, bamboo tubes of greens mixed with *marita* sauce, and bananas. The visitors are asked to stop dancing and gather around while a presentation speech is made by one of the men responsible for the invitation. As he slowly walks around and around the food that has been laid out in a number of piles, the speechmaker recounts the relations of the two groups: their mutual assistance in fighting, their exchange of women and wealth, their hospitality to each other in times of defeat. He then points out the piles of food being presented to each of the invited men. While the form of the presentation ceremony is corporate, with one or sometimes two men speaking for the entire local population about the relationship of the two local populations, the piles of food are presented by individuals to individuals. The recipients do not make speeches in return. When the speech of presentation is finished they gather their portions and distribute them to those men who came to help them dance, and to their women.

Dancing continues throughout the night around low fires on the dance ground or, in case of rain, in the two large houses at the edge of the dance ground. As the night progresses the formations of visitors and locals becomes more and more mixed as men frequently switch back and forth. By dawn almost everyone has danced with everyone else.

Many of the women and girls retire to women's houses long before dawn. Others, however, remain huddled around the fires, where they may watch the dancers and where they themselves remain under the surveillance of their own menfolk. Opportunities for direct contact with eligible members of the opposite sex of the other group are thus limited. Indirect contacts, however, are made. A visiting girl may talk freely with her cross-cousins, either male or female, among the locals, for instance, and may ask one of them to tell a local young man whom she finds appealing to come to court her on an appointed day. Local girls may not receive suitors from other local populations until the entire year-long *kaiko* is finished, but they may express their admiration for particular young men to their visiting kin, who may be depended upon to report their sentiments to the admired men.

On the other hand, it is said that men do not make direct overtures to women. Young men say that a rejection would be bruited about, making them objects of ridicule. A *yu wundi* (good man) —one who dances strongly, whose plume waves bravely, and whose adornment is rich—will attract women. It is for men to entice women, not to approach them.

At dawn the dancing slackens. The dancing ground now becomes a trading ground as men from the Jimi Valley, from across the Simbai River, and from up and down the valley offer their wares. Bird feathers, gold-lip shells, green sea snail shells, marsupial furs, axes, and bushknives are, in terms of value and the frequency of exchange, the most important items entering transactions. Baby pigs, cassowaries, and salt are occasionally traded, and a number of minor items, including pigments, tobacco, loose marsupial fur, green beetles, and orchid fiber waistbands are also offered. In recent years money, usually in the form of one shilling coins, but occasionally in one pound notes, has entered into transactions. During 1962 and 1963, however, it had not yet become a universal medium of exchange. While it could be exchanged for every kind of commodity, it was not always acceptable. If a man was firm in his desire to obtain a gold-lip shell for his Princess Stephanie plume he would not accept money for it. Money, in short, was traded like other items.

The transactions that take place on the dance ground are completed on the spot; a man both gives and receives at the same time. Moreover, the exchanges are impersonal. The relationship between the parties to a transaction, who may have never seen each other before, may last no longer than the time it takes to complete the trade. At the men's houses, however, a different kind of exchange takes place. Here men from other places give to their kinsmen or trading partners in the local group valuables for which they do not receive immediate return. Men from north of the Simbai River leave plumes with the Tsembaga, who will, after wearing them, exchange them with Jimi Valley men for shells or axes. Jimi Valley men, conversely, leave shells and axes with the Tsembaga, who will exchange them, at future *kaiko* or on future visits to the Simbai's northern banks, for plumes, or in earlier times, native salt. It is often months before such transactions are completed, that is, before a man receives a return for a valuable he has left with a trading partner.

Trading generally lasts for an hour or two. Toward the end some of the young men, although footsore and hoarse, begin to dance again, thus demonstrating their endurance to whatever spectators remain. The dancing is sporadic, however, and ceases altogether by mid-morning. By this time the guests, except those lingering on to visit a little longer with kinsmen, have departed for home.

THE KAIKO, WOMEN, AND GOODS

It is obvious that the supralocal *kaiko* assemblages facilitate mate selection and marriage by providing settings in which marriageable, or soon to be marriageable, girls and young men are brought together. At least seven of the twenty-eight wives and widows of other-than-Tsembaga origin living on Tsembaga territory during 1962–63 made their first overtures to the men they married after being attracted to them at a *kaiko*. It is also obvious that the *kaiko*, by providing market-like settings, facilitates the exchange of goods, some of which, particularly axes and salt, are necessary for survival.

The *kaiko* also facilitates the exchange of goods, and, perhaps, the movement of women, in a less direct and more subtle way. In the last chapter the supralocal exchange system was described. It was suggested that an exchange apparatus in which only two items, in this case salt and axes, critical to either metabolism or subsistence are exchangeable for each other might well be unworkable, since the production of each of the two items would be determined not by its own demand, but by the limited demand for the item for which it is exchanged. It was further suggested that the introduction into the exchange system of valuables, the demand for which is unlimited, provides a mechanism for regulating the production of each of the two critical items in accordance with their own demand.

Valuables, thus, although themselves nonutilitarian, are crucial to provisioning the population with utilitarian commodities.

That valuables should flow from a locality of greater population increase to a locality of lower population increase was also hypothesized. The accumulation of valuables in the locality of lower population increase might be used to obtain women from one of higher population increase, adjusting differences in population between the two localities and thus contributing to the long-term adjustment of population dispersal over the entire area.

The nonutilitarian valuables upon which the Maring exchange system depended most heavily in earlier times were bird feathers, shells, and "bridal" axes. In 1962 and 1963 the importance of shells and bird feathers remained undiminished. The demand for shells is created by their use as payments to affines. It is also the case that shells, being durable, may and do form part of the ordinary daily costume. Bridal axes were also in demand because they could be included in payments to affines, and although they did not form part of a man's daily accoutrements they were carried during visits to friendly groups. Informants say that to appear at the houses of another group bearing only a bridal ax indicated both friendship and reliance upon the host's hospitality, for such implements were hardly fit for either fighting or chopping wood. Valuable bird feathers, however, did not, except occasionally in the Jimi Valley, figure in affinal payments, nor, since they are perishable, did they form part of the daily or even the visiting costume.[11] Their use is confined almost exclusively to their display at the *kaiko*. It may thus be said that the *kaiko* contributes to the movement of critical commodities, and perhaps to the movement of women as well, by creating a demand for one class of valuables upon which the operation of the exchange system depends.

RITUAL AS DISPLAY BEHAVIOR

The term *ritual* has been taken in this study to refer to the performance of conventional acts explicitly directed toward the involvement of nonempirical or supernatural agencies in the affairs of the participants. Although they may have social, demographic, nutri-

11. Less valuable bird feathers may form part of daily or visiting dress. Men will frequently wear a single parrot or eagle feather, for instance.

tional, and ecological consequences, the events already described in this study—from the hanging of the fighting stones to the inauguration of *kaiko de*—are comfortably accommodated by this description, for each is undertaken explicitly to effect changes in the relationships of the participants with various categories of supernaturals. Together with other events that remain as yet undescribed these rituals form an articulated series through which the changes occur in a certain sequence. Because the relationships among the relevant natural and supernatural entities at the termination of the sequence are such as to permit, encourage, or even demand its repetition, the sequence may aptly be termed a ritual cycle.

Kaiko entertainments, like the other events, have a prescribed place in this cycle. They occur only after the uprooting of the *rumbim* and before certain other events that will be described later. But these entertainments, although they include addresses to the spirits, are not primarily directed toward the involvement of nonempirical or supernatural agencies in the affairs of the participants. The dominant and explicit concern of the participants lies in their relations with other participants. Although the entertainments are intrinsic elements in a series that may in its totality be oriented toward the supernatural, their explicit aims are secular. But the term *ritual* is not restricted in its application to events involving the invocation of supernatural or nonempirical agencies. Indeed, the stereotypic or conventional aspects of acts are likely to be more fundamental than their sacred or supernatural aspects in identifying them as rituals, and both anthropologists (e.g., Goffman, 1956:478 passim, and Leach, 1954:10ff) as well as ethologists (e.g., Blest, 1961; Elkin, 1963; Hinde and Tinbergen, 1958; Tinbergen, 1952, 1963) have used the term to designate certain events, occurring among both men and other animals, in which one or more participants transmit, through conventional sign or symbol, information concerning their own physiological, psychological, or sociological states to other participants. They have used the term *ritual,* in other words, to refer to a class of communication events.

There are, of course, many kinds of communication events, and no purpose is served by regarding them all as ritual. But ritual may be distinguished from other modes of communication by its special language, which is conventional display. Put in terms of communications models, if ritual is regarded as a channel, conventional display is the code appropriate to that channel.

Kaiko entertainments, although explicitly secular in intention, may be regarded as rituals not only because their most obvious characteristic is conventional display, but because the displays communicate among participants certain information that, given other aspects of Maring culture, can hardly be communicated in any other way.

The notion of display behavior has already been introduced several times in this study. It has been suggested the "small" or "nothing" fight may be regarded as agonistic territorial display, similar to that which has been observed among species other than man, rather than as sanguinary fighting. It was also suggested that agonistic display is one aspect of the stake-planting ritual, which demarcates or ratifies the territorial boundary. The messages transmitted through such displays are, obviously, ones of threat. Other messages are transmitted by the displays in which friendly groups participate at *kaiko* entertainments.

These displays have two main aspects. The first of these, following V. C. Wynne-Edwards, may be termed *epigamic.* Speaking of nonhuman species Wynne-Edwards uses this term to refer to "displays that characterize the marital relations of the sexes and typically culminate in fertilisation" (1962:17). They are in other words amatory displays, which form all or part of a courtship procedure. The dancing of males at a *kaiko* constitutes the first phase of a conventionalized courtship procedure that, if the female spectators respond favorably, may continue for some of the participants in other contexts.

Certain information is imparted by the massed dancing of the males. First, it presents to the female spectators larger samples of the males of unfamiliar local groups than they are likely to see assembled at any other time or place. The males, fruthermore, signal by their participation in the dance their general interest in the females as a class. It would be difficult to conceive a more economical means for communicating information concerning the availability of males than the sample presentation of the dance. Through it females are able, on a single occasion, to gain some familiarity with all or most of the eligible males of local populations in which they themselves are not resident and which they may visit only rarely.

The amatory display of the males does more than present a sample to the females, however. It also provides them with a basis for differentiating among the males. The appearance of the individual men—their dancing and the richness of their adornment—indicates to females their strength or endurance and their wealth or the wealth of their connections.

Wynne-Edwards suggests that such displays in the animal kingdom are selective mechanisms, since "individuals that were undernourished or depressed would presumably have greater difficulty in achieving mating than the dominant and well-fed" (1962:251). Whether or not the choices that Maring females make are usually responses to the comparative quality of the performances of the individual males cannot be answered here, for the data are insufficient. A number of men maintain, however, that their wives were first attracted to them because of the admirable figures they cut in the dance. It may be pointed out, moreover, that characteristics displayed in dancing are not irrelevant to more prosaic activities. Endurance or strength is as vital to gardening and fighting as to dancing, and the richness of a man's adornment gives some indication of his ability to pay for the woman whom he may attract. This information, it may be added, is not only communicated to the eligible females, but also to their male agnates, whose attitudes toward suitors, although sometimes ignored, are germane to all matings.

It is only the behavior of the men that may be characterized as amatory display, but of course it is also the case that the marriageable women of the hosts and guests are made available for inspection by men at *kaiko* entertainments. While the display behavior of the dancing men may elicit overtures from the women, the presence of the women may also prompt the men to make overtures to the male kinsmen of the women. A man seeking a wife for himself, his son, or his younger brother may approach the father or brothers of a girl who attracts him and for whom he is either able to exchange a sister or daughter or make substantial payments.

The *kaiko*, in short, forms part of two procedures that facilitate sexual pairing. First, by providing a setting in which massed amatory display takes place, it offers an opportunity for females to invite courtship from specific individuals among a large and perhaps previously unfamiliar sample of available males. Second, it is an occasion upon which large numbers of females are presented for inspection, thus providing a basis for negotiations between males concerning the disposal of females. The two procedures, one involving selection by females, the other by males, sometimes conflict, for girls' choices do not always coincide with those of their fathers or brothers, but this need not be considered here.

Following Wynne-Edwards (1962:16), the term *epideictic* may be applied to the second major aspect of displays at *kaiko* entertainments. Epideictic displays are those that impart to the participants information concerning the population's size or density prior to behavior that may affect that size or density. Included by Wynne-Edwards are the "dancing of gnats and midges, the milling of whirligig-beetles, the manoeuvres of birds and bats at roosting time, the choruses of birds, bats, frogs, insects and shrimps" (1962: 16). Epideictic displays usually occur at conventional times and frequently at "traditional places" (1962:17).

The specification by Wynne-Edwards that such displays precede events that "restore or shift the balance of population" justifies the use of the term *epideictic* in reference to *kaiko* entertainments. One of the ways of restoring or shifting the balance of population is by adjusting the dispersion of organisms over the land. It must not be forgotten that the occurrence of the *kaiko* immediately precedes the termination of the truce. After the *kaiko* the existing pattern of population dispersion may again be tested through renewed hostilities. In anticipation of renewed hostilities it is important for the members of a local population to assess the extent to which it will be supported by its allies. Among the Maring it is not possible to base such assessments upon promises of support from authoritative political leaders, men who can command the performance of others, for such do not exist. The decision to participate in fighting as an ally is at the discretion of each individual male.

The Tsembaga say that "those who come to our *kaiko* will also come to our fights." This native interpretation of *kaiko* attendance is also given expression by an invited group. Preparations for departure to a *kaiko* at another place include ritual performances similar to those that precede a fight. Fight packages are applied to the heads and hearts of the dancers and *gir* to their feet so that they will dance strongly, just as, during warfare, they are applied so that they will fight strongly. It has already been indicated that these acts have amatory aspects, but it is also said that dancing is like fighting. The visitors' procession is led by men carrying fight packages, and their entrance upon the dance ground of their hosts is martial. To join a group in dancing is the symbolic expression of willingness to join them in fighting.

The size of a visiting dancing contingent is a product of many factors. Most important of these are the size of the local population to which the invitation has been extended, the number of kin or formal trading connections between hosts and guests, and the extent to which the formally invited men can induce others to support them. The last is itself a product of the relations of the formally invited men to members of their own and other groups.

These are also the factors that most importantly affect the recruitment of allies in time of warfare. Mobilization to attend a *kaiko* thus exercises the connections through which mobilization for warfare is accomplished, and the size of the dancing contingent signals the total strength and effectiveness of these connections. The hosts, thus, can base assessments of the extent to which friendly groups will support their belligerent enterprises upon the samples that are presented to them in the form of dancing contingents. Given the absence of authoritative political leaders, it is difficult to imagine how this information could be economically communicated without display or some other means for presenting a sample.

It is true, of course, that participation in a dance is different from participation in a fight, and men who may be pleased to attend the former may be reluctant to engage in the latter. Display accommodates deception and dissembling, but so does language and the more specialized codes that depend upon language. Indeed, the ability to transmit lies is common to all means of communication that employ symbols. In this connection, however, it should be recalled that although *kaiko* entertainments are essentially secular, visitors do address their ancestors before arriving, and that these addresses and the acts accompanying them are similar to those that are undertaken prior to assisting another group in warfare. It may be suggested that the involvement of the spirits in the *kaiko* participation of the visitors sanctifies the information they transmit to their hosts through display. The sanctification of messages may be of importance in a communication system that can easily accommodate falsehood: on the one hand men may be loath to sanctify information they do not take seriously, and on the other hand statements that have been sanctified may be more credible to recipients than mere promises.

In addition to the epigamic and epideictic messages, display at *kaiko* entertainments transmits other information. Before the food presentation, for instance, a formally invited man watches his host assemble his portion, thus learning how many men his host has been able to induce to assist him in this effort. The invited man's followers, from the size of their portions, may estimate the strength of the connections between the host and the man whom they are "helping to dance." The hosts, at the same time, may assess the influence of the formally invited men by observing the number of people to whom each redistributes the food presented him. The displays thus transmit information concerning not only the strength or size but also the structure of the participating groups.

The Culmination of the Kaiko

With the coming of the "dry" season, in May or June on Tsembaga territory, new gardens are cut. Ordinarily, the greater part of the men's work would be finished by late August, but in 1963 unseasonable rains, falling mainly in the daytime, and heavy overcast seriously impeded burning. Preparations for the *konj kaiko*, the "pig *kaiko*," the event culminating not only the year-long festival but also the entire ritual cycle, were nevertheless initiated by some men in mid-August, and by early September most other men had followed them.

TRAPPING EELS

As preparations for uprooting the *rumbim* begin with the trapping of *ma*, the "pigs of the *rawa mugi*," so preparations for the *konj kaiko* begin with the trapping of eels, the "pigs of the *raua mai*."

The minimal agnatic groups, either clans or subclans, form the trapping units, as they did for the trapping of *ma*, and similar to the trapping of *ma*, the locations in which each agnatic group may place its traps are restricted. Although at other times men may place traps anywhere, they now must set them in traditional places associated with their own spirits of the low ground, for it is only with them that pigs may be exchanged.

The personnel participating in the trapping of eels differs to some extent from those involved in taking *ma*. Fight magic men, who remain for life dedicated to the *rawa mugi*, are precluded from even touching with their hands the cold, wet eels, and they therefore take no part in trapping them.

The prohibitions burdening the men engaged in eel trapping are similar to those under which they labored when trapping *ma*. Now, however, it is the high ground, the *kamuŋga*, which is to be avoided, and the people suffer no prohibition against *marita*.

As the flesh of the trapped *ma* is preserved, so, in a sense, is that of the eels. After being taken from the traps they are kept alive in individual cylindrical bark cages left submerged in the streams. Although they are not fed, some of the captured fish maintain life for as long as two months, apparently subsisting on whatever bits of plant and animal material is carried to them by the current.

As in the case of *ma* and pigs, there is no particular number of eels required for the performance of the rituals in which they figure. A sufficiency of eels seems to be defined, rather, by the toleration of these fish for captivity. As the trapping period continues, the trappers suffer mounting losses among the imprisoned eels, due perhaps to possible lowered food intake and lack of exercise. Sometimes, too, in the heavy run-off of water following downpours, both traps and cages are carried away. With each loss there is increased talk of getting on with the remaining preparations for the *konj kaiko* before all of the eels are lost.

PREPARATIONS AT THE RAKU AND DANCE GROUND

By early October 1963 some men, not waiting for the "talk to become one," after being advised by shamans of the wishes of the ancestors, began to make preparations both at designated *raku* and on the dance ground for the *konj kaiko,* the event that brings the festival and the cycle to a close.

On the dance ground repairs were made to the two houses, and late in the middle of the month a ceremonial fence, called the *pave*, was built on the slope above one end of the dance ground. Constructed of saplings and covered with foliage, this light three-sided structure, about 15′ high, enclosed an area of about 30′x50′. This area was increased by the incorporation into it of the large men's house enclosure that stood directly above it. The longest dimension of the *pave*, that facing the dance ground, was broken near its center by a single window about 1′ square and approximately 4′ above the ground.

By mid-October work had started at the *raku* of all of the minimal agnatic units. These included sites where the residences of men slain in the last fight had stood, for it was these men specifically to whom pigs were to be dedicated. These *raku* were in some, but not all, instances separate from those at which pigs were to be killed for the spirits of the low ground. Most of the latter were to be killed at *raku* that had always been used for this purpose. If the slain men had lived near these traditional sites a single *raku* might be used for both categories of ancestors.

Anomalies in the general pattern of a separate *raku* for each of the minimal agnatic units gave, perhaps, some evidence of past or on-going changes in agnatic structure and residence patterns. Three of the adult males of the Wendekai subclan of the Merkai clan prepared a *raku* separate from that of the other eight adult males. The three had been separated by food and fire taboos from their subclan brothers, but such taboos, it has already been mentioned, split other groups that did maintain the use of single *raku*. Some members of the Atigai subclan of the Tsembaga clan, on the other hand, joined with the Atigai subclan of the Kamuŋgagai clan, as had their fathers, who, they said, had lived sororilocally with the Kamuŋgagai Atigai. The remaining Tsembaga Atigai shared a *raku* with the Tsembaga Wendekai subclan and the unsegmented Tomegai clan.

Early preparations at both kinds of *raku* included the clearing of underbrush and the erection of roofed but unwalled structures for the storage of firewood and vegetables, and for shelter from rain. Additional structures were erected at those *raku* where pigs were to be killed for the spirits of the low ground. These were *timbi* houses, named after the trees that formed the center posts (*Myrtaceae*, ? *Cleistanthus* sp.), which, frequently found by wide still places in streams, are said to be *koipa maŋgiaŋ*'s own trees. *Timbi* houses are round, six to nine feet in diameter, and the *timbi* center posts, with some leaves remaining, project several feet above the conical roof.

It was in these houses that the eels were to be cooked with pig bellies in one or two earth ovens, depending upon whether food taboos split the group.

The erection of the center post is accompanied by a brief ritual, the object of which is fertility and abundance. A sleeping mat is laid upon the earth floor of the still roofless *timbi* house, and valuables to be offered to *koipa maŋgiaŋ*, "he who gives us eels," are spread upon it. The butt end of the center post is placed upon the mat, and one of the older men, singing in a soft falsetto voice, first cleans moss off its bark with a bamboo scrapper then, with an ax, marks its length with a zig-zag line, representing, informants say, an eel. The debris falls upon the wealth objects, and the song is concerned with the increase of wealth and the thoughts of trading partners. "Let him think of me and send one gold-lip shell. Let him think of me and send one ax," etc. When the work is finished, the moss and bark are gathered up to be cooked with vegetables and eaten by all but fight magic men to enhance both fertility and growth.

The spirits of rot, the fathers and grandfathers of those present who died from causes other than violence, are then addressed by name. They are thanked for the eels and told that they are now being offered valuables at this place where they, when living, killed pigs, and where those presently living would soon kill more. They are told to accept the valuables and give some to *koipa maŋgiaŋ*, and are asked to look after the women and children, each of whose names are mentioned. The red spirits are then addressed. They are reminded that wealth was previously given to them but that now it is being given to "those who gave us eels." They are asked, however, to continue to look out for the men, all of whom are named. All the men and boys then place their hands on the *timbi* center post as it is thrust in the ground. The shells and beads are then hung from a low branch left on the center post for this purpose, and the axes are planted in a circle around the base of the post. After the roof is completed, the valuables are returned to their owners.

As in the case of the planting of *rumbim* and *amame*, the symbolism of the ritual paraphernalia, particularly the *timbi* house itself, seems sexual. Of greater interest is the further adjustment in the relations of the participants with both categories of ancestors. With the hanging of the fighting stones and the taking of the *riŋgi*, the dominant relationship was with the red spirits. This was modified when *rumbim* and *amame* were planted, but a heavy debt to both categories of ancestors was acknowledged. When these plants were uprooted a further adjustment between the living and the two categories of spirits was expressed in the abrogation of most of the taboos remaining from the time of the fight, in the reestablishment of reciprocity with the red spirits, and, perhaps, in a commuion with the spirits of the low ground. A further adjustment takes place when the *timbi* center post is set. Reciprocity, to be bound by a forthcoming "pig exchange," is being reestablished with the spirits of the low ground. However, it is important to note that the red spirits are also addressed during the planting of the *timbi*. The Tsembaga say that if they weren't they might grow jealous and desert the living, leaving them open to slaughter in the next round of fights. The goal of the rituals through which the Tsembaga and other Marings proceed does not seem to be the replacement of the red spirits by the spirits of the low ground, but rather a redefinition of the balance in the relationships of the living with the two categories of supernaturals.

THE EFFECT OF THE PLANS OF OTHER GROUPS ON THE TIMING OF THE KONJ KAIKO

The plans of other groups sometimes affect the timing of the *konj kaiko*. The Tuguma uprooted their *rumbim* when their *pengup* variety *marita* ripened in October 1963, one year later than the Tsembaga's. It has already been mentioned that men of local populations who have *rumbim* in the ground may not eat the flesh of pigs killed in connection with the *kaiko* of other groups. The Tsembaga, who might otherwise have been able to stage their *konj kaiko* several weeks earlier, were forced to wait until their most important allies and closest neighbors had uprooted their *rumbim*, thus becoming free to eat the pork to be presented to them.

Delays in terminating *kaiko* for this reason must have been common if not the rule, considering the frequency of fighting throughout the Maring area and the number of groups from which allies were usually drawn. Their consequences in recent years have been trivial. The Tsembaga complained only about increasing deaths among their captured eels. Before pacification, however, they may have had more important consequences. In situations in which the *kaiko* of a pair of enemies were separated by one year, such delays may have served to eliminate entirely any period during which one was free to attack while the other was not.

THE MAMP GUNČ

When all but the final preparations at the *raku* are completed some of the young men undergo ritual dedication to the red spirits. The initiates are secluded in the men's house enclosure behind the *pave* and their hair is worked into constructions called *mamp gunč*. Round frames, about six inches high, made of the bark of the *kirim* tree (*Lauraceae* sp.) are placed on their heads like crowns. The hair, which has remained uncut since puberty, is pulled up through the center and down over the sides of the frame, hiding it completely. The melted sap of an unidentified tree called *gunč*, which gives the construction its name ("head *gunč*"), and which, upon cooling, leaves the surfaces hard, is then applied. Finally, the headdress is dyed red, with trade pigments now.

Only a few of the fight magic men possess the skill of making *mamp gunč*, and the procedure is protracted. The arrangement of the hair over the frame takes almost a day for each novice, and the application of the melted *gunč*, bit by bit with an arrow point, takes another day. It is also highly ritualized, but the details of the rituals need not be discussed here. It is sufficient to say that the red spirits are told that the boys are putting on these "red things" for them and their help is asked in making them shapely, hard, and bright. All of the remaining ritual procedures have the same goal.

As might be expected of anything associated with the red spirits, the young men assume a number of stringent taboos along with their *mamp gunč*. These proscribe, among other things, sexual intercourse and visiting the *wora* while the *mamp gunč* remain attached to their heads. More notable, however, are absolute prohibitions against drinking water, chewing sugar cane, and eating cucumber, paw paw, and hibiscus leaves. These prohibitions lasted, in the case of the Tsembaga boys, for eight days. They did not exert themselves during this period, and the roots and greens they were allowed to eat evidently contained sufficient moisture to prevent serious dehydration. Some violations of the taboos did occur, however. I know personally of two instances in which *gunč yu*, as young men wearing *mamp gunč* are called, took liquids. One drank a few sips of coffee, and the other, after complaining of a cough, took cough syrup with a water chaser. Both of these exotic liquids were available only from us and were so uncommon and unfamiliar that the fight magic men had not thought to proscribe them.

When work on their headdresses is finished, the new *mamp gunč* men are "brought out"; that is, they end their seclusion by dancing publicly for the local women and whatever visitors from other places care to watch.

It has already been mentioned that only some of the Tsembaga men took *mamp gunč* at the 1963 *kaiko.* In fact, only five did so. This was an unusually small number, but it seems to be the case that seldom, if ever, did all young men put on *mamp gunč* at the same time.

Mamp gunč, first, are associated with the killing of enemies. Some informants say that *mamp gunč* may be worn only by young men whose agnates have killed enemies in the last fight. Others say that if a member of the subterritorial group has killed, they may be worn, while yet others maintain that the *mamp gunč* may be put on if any member of the local population has slain an enemy. It is not possible to derive any rule from what went on among the Tsembaga in 1963, for the proceedings were unusual. The five young men who put on *mamp gunč* either had taken refuge among the Kauwasi or were agnates of those who had. They justified their *mamp gunč* on the basis of the slaying, during the Kauwasi-Monambant fight, of a Kundagai allied to the Monambant. The Tsembaga who took refuge in Monambant, however, said that they had killed no Kundagai and would await the Monambant *kaiko* to wear *mamp gunč,* for they had helped Monambant kill Kauwasi.

Even in situations in which all young men are related to killers in such a manner as to be eligible to put on a *mamp gunč,* it is unlikely that all would do so. It is said that two brothers, for instance, cannot take *mamp gunč* at the same time, for a man who is wearing a *mamp gunč* cannot assist with the slaughter of the pigs, the gathering of firewood and vegetables, and the other tasks associated with the *konj kaiko.* Those who do not take the *mamp gunč* at the *kaiko* of their own local population, however, may put them on at the *kaiko* of other local populations to which one's own has been allied. To qualify to put on a *mamp gunč* at the *kaiko* of an allied group it seems sufficient to belong to the same local population as any man who has participated in the killing of one of that group's enemies. It is not necessary to be an agnate of the killer.

It is interesting to note that the young men who join in putting on the *mamp gunč* may, although they seldom do, refer to each other as "brother." It could be said that the *mamp gunč* procedure is a device producing supralocal age classes. The political and structural potentialities of such age classes, if they may be so termed, remain unutilized, however.

Another kind of group may also be crystallized by, or at least become apparent in, the *mamp gunč* rituals. While there are at the time of most *kaiko* probably many young men who are eligible to put on *mamp gunč*, there are always certain of them who must do so. These are the young men who have been designated by each of the fight magic men to be heirs to their ritual knowledge. Whenever possible they are the biological sons of the fight magic men, but if a fight magic man has no son he will designate a brother's son, the sons of true brothers being selected before the sons of classificatory brothers.

If a priest is one who performs prescribed rituals at specified times for the benefit of a congregation, it may be said that fight magic men are priests. The putting on of the *mamp gunč*, it may further be said, specifies a group of young men who will inherit priesthood. Hereditary priesthoods, even more than age classes, have served as foci of political activity in many societies. Among the Maring, however, the extent to which ritual knowledge has been converted into secular authority is minimal. Continuity of ritual knowledge is maintained, but the political possibilities of hereditary priesthood remains, like that of age classing, potential. It is beyond the scope of this study to discuss the reasons underlying the lack of development of either age classes or heredity as bases of political organization. It may be suggested, however, that the ritual cycle itself defines sufficiently the tasks that require aggregated effort, and that in groups as small as those of Maring local populations the efforts of entire groups may be mobilized without recourse to the formal structures or positions that could be yielded by age classes or hereditary priesthoods.

FINAL PREPARATIONS

In the last days before the *konj kaiko*, activity reaches a high pitch. Above-ground ovens must be built for pigs to be dedicated to the red spirits and firewood must be gathered at the *raku.* Women must gather large quantities of the ferns and other greens with which pigs are cooked. Many of the visitors who were to receive pork arrived in advance, for they are expected to help in the preparations. It is also necessary to assemble captured eels at locations convenient to the *raku,* so they were deposited in their cages in nearby streams. New paths were cut from their places of deposition to the *raku,* and at their termini light archways were constructed.

Before the pigs are killed, the ground must again be rid of any corruption deriving from spirits or sorcerers. To this end, two days before the *konj kaiko* a number of fight magic men gathered on the government walking track, where they bespelled large bamboos full of water. These were distributed to representatives of each of the minimal agnatic units to use at the *raku* and on the paths. While sprinkling the contents of the bamboos over the ground the men recounted, to both the living and the dead, the story of their defeat and departure and the subsequent pollution of the ground by Kundagai spirits and sorcerers, and announced that they were now counteracting this pollution with magical water.

THE ABROGATION OF TABOOS

On the day before the major pig slaughter fourteen *aček konj,* "taboo pigs," were sacrificed. Although both categories of ancestors were told in the addresses before the killing that the living wished to terminate taboos, the slaughtered animals were cooked in above-ground ovens, for the taboos that applied most generally, those arising from warfare, were associated with the red spirits.

With the opening of the oven each individual brushed himself with the *gañiŋgai* shrub, made the spitting sound, and announced separately the taboos he was terminating. The pig was then eaten by those who had participated in the ritual. The variety of taboos abrogated was great, and it is convenient to separate them into three classes: those associated with mourning, intralocal disputes, and warfare.

During mourning a woman is separated by fire taboos (*aček*) from nonmourning women and from men, because the bones of the deceased remain in her house, by taboos against touching or even conversation. If the deceased was a man, moreover, the woman is burdened with all of the food taboos he suffered when alive.

Just before the *konj kaiko* women who have been keeping the bones of deceased persons at their houses bury the remains at the *raku,* so that the spirits of these deceased may partake of the flesh of the pigs soon to be killed. Burying the bones at the *raku* effectively terminates the mourning period, and the taboos associated with it are abrogated with the killing of the taboo pigs.

Mourning taboos apply to people other than the women who look after the bones. It is customary for the deceased's close kin (members of his families of orientation and procreation) to give up, as an expression of sorrow and loss, one of the deceased's favorite varieties of each of the major categories of foodstuffs (i.e., one variety of taro, one of banana, etc.). These voluntary renunciations are termed *moi.* If the deceased was a woman, moreover, the men whose pigs she looked after customarily give up the consumption of pork for the duration of the mourning period, and widowers also often announce that they will neither marry nor engage in sexual intercourse for an indefinite period. These restrictions are also abrogated with the killing of the taboo pig.

Antagonisms between members of the group arising out of serious arguments, disputes, and hurts are given symbolic expression in the form of interdining and food-sharing taboos also known as "*aček.*" The principals refuse to eat food cooked over the same fire, and each refuses to eat food grown by the other. Taboos arising out of a variety of incidents ranging from obscene name calling to homicide were terminated with the killing of the taboo pigs.

Some of the taboos arising out of warfare (also referred to as "*aček*") are also terminated with the killing of the taboo pigs. To abrogate the full range of taboos applying to intercourse between the agnatic groups of the slayers and the slain takes four generations. Children of the opposing principals may neither eat foodstuffs raised by their opposites nor share a cooking fire. Grandchildren may eat food grown by each other, but may not eat food cooked over the same fire. Great-grandchildren may, after renouncing the taboo, share the same cooking fire. Individuals who are related through wives or mothers to erstwhile enemies constitute an exception to the general applicability of this rule, but this will be discussed later in another context.

In 1963, young people of the Kamuŋgagai clan terminated the

taboo on food grown by the Kekai clan, who had killed a member of their grandparental generation during warfare perhaps fifty years earlier. On the same day, men of the Merkai clan renounced a taboo upon the fires of the Raweŋ clan of Tuguma, who had killed two members of their great-grandparental generation so long ago that no one could provide information concerning the circumstances surrounding the deaths.

The Tsembaga also terminated some of the restrictions on intercourse with their enemies of the last fight, the Kundagai, with the sacrifice of the taboo pigs. These were the taboos on walking through Kundagai territory, talking to and touching Kundagai, and visiting (but not entering) Kundagai residences.

The abrogation of taboos on intercourse with the Kundagai at this time was irregular. In earlier times the restrictions would have remained in effect until peace-making ceremonies could be conducted some time in the future, but the establishment of a Simbai patrol post and the administration policy of drawing people into control by requiring of them corporate effort made the abrogation of these taboos ex post facto as it was. While carrying cargo for government patrols, for instance, most Tsembaga had crossed Kundagai territory, as most Kundagai had crossed theirs, and in building a bridge ordered by the government across the stream separating their territories members of the two groups had already been forced into cooperation.

It has already been mentioned that taboos arising out of the warfare of other groups may split local populations. Those Tsembaga who had fought on opposite sides of the Monambant-Kauwasi fight terminated their taboo on each other's foodstuffs with the killing of the taboo pigs. The taboo on each other's fires, however, was to remain in effect until the Monambant terminated their *kaiko* in 1964.

TABOOS, SOCIAL CONTROL, AND INTERGROUP RELATIONS

Taboos on interdining, food sharing, and on other social relations are a pervasive aspect of Maring social and political life. Interpretations of the nature and function of ritual advanced by Bateson (1936), Freud (1907), Gluckman (1962), and Reik (1947), although they cannot be tested, suggest certain ideas concerning the role of taboo in these relations. The extension of interpretations of ritual to an examination of taboos is warranted because taboos, which may be defined as supernaturally sanctioned proscriptions of physically feasible behavior, form a logical pair with ritual; indeed, they may be regarded as "negative rituals." Among the Maring, moreover, the relationship is not only logical but actual, for taboos are also both ritually assumed and ritually abrogated.

Gluckman has suggested that "social rules and values, established by diverse relations, themselves move individuals and sub-groups to dispute with their fellows in their main group of allegiance," and that "ritual operates to cloak the fundamental conflicts which are set up" (1962:40). He does not suggest that rituals settle conflicts. "The whole point of the analysis is that they cannot do so" (1962:46). He suggests instead that conflicts between parties may be concealed by rituals, which "in fact may lead to temporary truces." The ability of ritual to perform such a function, Freud's writings suggest, may lie in the nature of ritual itself. In remarking upon the similarity between the obsessional ceremonials of neurotics and the religious rituals of the pious, Freud notes that both represent compromises between opposing forces. At one and the same time they both suppress and give expression to feelings that may be dangerous to an individual or to a group.

These formulations suggest that the taboos that prevail within Maring local groups are an important means of social control. Among the Maring these taboos arise out of antagonisms. I have already mentioned the basis of some of the antagonisms giving rise to the taboos prevailing among the Tsembaga in 1963. Some were generated by events that are the frequent concomitants of coresidence: insult, assault, woman-stealing, and so on. Others arose out of what Gluckman would call the fundamental constitution of Maring society, that is, out of diverse affiliations and conflicting responsibilities. Thus, taboos to which all Tsembaga were subject resulted from the Monambant-Kauwasi war of 1955. Maring rules of participation in warfare are such that some Tsembaga men could not avoid fighting on each side and thus fighting each other. The Tsembaga were in exile at the time of the war, and those who were living with the Monambant and Kauwasi had to join, as principal combatants, their hosts, who had provided them with land and shelter, and two Tsembaga men were killed and several were wounded in the fighting. When, however, the Tsembaga exile ended and they reassembled on their own territory it was necessary for those who had fought against each other in the Monambant-Kauwasi war to become again members of a single cooperating and coresidential group.

It may be suggested, following Freud and Gluckman, that taboos played an important part in this reamalgamation. The taboos served to define areas of behavior in which the anger and bitterness generated by death and injury could be expressed, while permitting cooperation in most of the important tasks of living. In Freud's terms the taboos represented a compromise between the needs to express and to suppress socially dangerous feelings. They required or allowed antagonism to be stated frequently and formally in controlled circumstances: by men cooking their food side by side, but over separate fires; by the refusal of one man to enter another's house; by the refusal to eat food grown by another. It may be suggested that the frequent but relatively harmless statement of antagonism in areas of behavior narrowly defined by taboos inhibited their more generalized, less predictable, and therefore more dangerous, expression. The supernatural nature of the taboos, moreover, may have relieved tension between antagonists by transferring animus from a mundane to a supernatural plane. These effects of observing taboos, I think, prevented ill-feeling from contaminating all aspects of the relationships of parties between whom grievance lay and thus permitted their cooperation in important tasks. It was in fact the case that men who refused to eat each others' food or at each others' fires helped each other in forest clearing and hunting, granted land to each other, assisted each other with affinal payments, and would have, had the need arisen, fought side by side in defense of the common territory.

It is interesting to note that it is to taboo rather than to ritual that the task of expressing and suppressing antagonism is assigned. I suggest that taboo is better fitted to the purpose than ritual, for the very act of expressing antagonism by observing a taboo turns the principals *away* from each other. Antagonism is thus stated through a means that avoids potentially dangerous confrontations.

This formulation is in some respects applicable to the relations between hostile local populations as well as to interpersonal relations within local populations. The ritual truce inaugurated by the planting of *rumbim* is reinforced by prohibitions on all intercourse between the hostile groups. These taboos, while expressing the hostility symbolically, minimize the possibility of actual truce violations. Although no active or conscious cooperation between the groups is facilitated or even allowed by these taboos, they at least make it possible for the members of each group to live in some security.

However, although taboos may play an important part in maintaining order within the local population, obviously there must be some means of abrogating them, otherwise their accumulation would eventually damage the coresidential and cooperative structure they help to maintain.

The renunciation of taboos involves the killing of pigs. Such slaughter, however, awaits the accumulation of a parasitic or competitive quantity of animals. To sacrifice animals sooner for the purpose of terminating taboos among members of the local group would diminish the quantity of pigs available for emergency and misfortune and it would delay the *kaiko*. Therefore it awaits the *kaiko,* when people can afford to put pigs to this purpose. The expressed attitudes and behavior of the Tsembaga suggest that by this time the taboos have fulfilled their function. The antagonisms that gave rise to them have cooled considerably and have been replaced by annoyance with the taboos themselves. Informants said that by the time of the *kaiko* they no longer bore any animus toward those with whom their relations were restricted by taboos, but continued to observe the taboos merely because their ancestors would punish them if they did not. With the renunciation of taboos people who had long been separated by them frequently embraced.

The *kaiko,* then, which is part of a mechanism for regulating relations between groups, may also be part of a mechanism, which includes taboos, for regulating relations between members of the same local group.

PIGS, EELS, AND FERTILITY

On the day following the abrogation of taboos, eighty-two more pigs were killed. Of these, fourteen were dedicated to the red spirits, one for each man killed in the last fight, and sixty-eight to the spirits of the low ground.

At former *kaiko,* informants say, the red spirits were told, when pigs were offered to them, that their deaths would soon be avenged. In 1963, however, they were told that because of the new presence of the government it was now impossible to avenge their deaths by natural means, and they were asked to eat the pig offered them so that they themselves might become strong enough to avenge their own deaths by visiting illness upon their slayers.

Addresses to the spirits of the low ground recounted the story of the defeat, the subsequent depredation of the land and desecration of the *raku* by the Kundagai, and finally, the return of the people to their own territory. Specifically named ancestors who had been invited to other places to receive sacrifices of pork while the living were in exile were now asked to return to their traditional places on their own territory to receive pig. The spirits as well as the living were thus resettled upon the territory they had previously abandoned.

All pigs were butchered and cooked at the *raku* shortly after their slaughter. Men, usually affines, from other places who were to receive entire pigs were at the *raku* to cook them themselves. The presentations were unceremonious. No speeches of either presentation or acceptance were made. The recipients, after giving their benefactors the return gift of an ax, bushknife, or shell,[12] simply proceeded to butcher and cook the animals presented them. Those recipients who had been allies in the last fight, however, returned the bellies of their animals to the donors to be salted and publicly presented to them on the following day. Men who were to receive less than an entire animal did not come to the *raku*. They received their portions on the following day already cooked.

At each *raku* at least one, and sometimes two, female pigs dedicated to the spirits of the low ground were further designated "*koipa* pigs." These animals played a part in a ritual the Tsembaga regard to be of great importance. As the *koipa* pigs were killed, the names of specific spirits of rot were called out, and they were asked to take this pig and pass it on to *koipa maŋgiaŋ*, "the man who gives us eels." After the *koipa* pigs were killed, the eels were brought from nearby streams by processions of young men and boys who carried the cages, decorated with *timbi* leaves, along the newly cleared paths. Women and girls awaited them at the light archways, made of *timbi* and other plants, at the end of these "roads of *koipa maŋgiaŋ*" on the edges of the *raku*. The men, women, boys, and girls then pro-

12. These presentations cannot be regarded as payment in any strict sense. A single wealth object may be traded for an infant animal. A full-grown pig is worth many times as much. It is not possible to be precise, since full-grown pigs are not traded.

ceeded together to the bodies of the slain *koipa* pigs, where the eels were removed from their cages. Holding them by their tails, women, children, and unmarried young men flailed the dead pigs with the eels until the eels too were dead. The bodies of the dead pigs were then rubbed with the eels. Informants say that the shoulders of the participants should also be rubbed with the eels, but I did not see this done.

The eels were then hung, along with shell valuables and beads, on the center post of the *timbi* houses, so that wealth might increase. Later in the day the valuables were returned to their owners, and as night fell the bellies of the *koipa* pigs were placed in the earth ovens inside the *timbi* houses along with the eels. When the ovens were opened in the morning, if the food was well cooked (as it invariably was), it was said that *koipa maŋgiaŋ* had come in the night and eaten, and that the people might now partake of the oven's contents so that they, their pigs, and their gardens would be fruitful. The spirits of rot were addressed first, being told that the living have given them many pigs, and now that they had eaten their fill they might return to their houses, the large trees in the low-altitude primary forest. They were asked to care for crops, pigs, and people so that these might be fertile and fast growing and were promised that when the crops came up they would be fed. They were finally told to take the contents of the ovens and give them to *koipa maŋgiaŋ*. Everyone, with the exception of fight magic men, ended their taboo on the consumption of eels by eating from this oven.

The purpose of this ritual is made explicit both in the addresses to spirits and in the statements of informants, all of whom agree that the performance is undertaken to ensure the fertility and growth of people, pigs, and gardens and the increase of wealth in the coming years. It is significant that this ritual occurs more or less concurrently with the final fulfillment of obligations incurred during the ritual cycle just ending, and with the termination of the taboos associated with these obligations. The slate has been, or is being, wiped clean. The local population now stands, or will shortly stand, in a relationship of equivalence or reciprocity rather than obligation to the spirits and allies, and a new cycle is beginning.[13]

13. It has not been the purpose of this study to discuss the psychic symbolism that may underlie Maring ritual objects or performances. A brief excursion into this

A Maring ritual cycle may be regarded as a series of adjustments in the relations of the people to two generally antithetic sets of spirits. The dominance of the red spirits is continually tempered as the debtor relationship is transformed into one of reciprocity. With the lowering of the fighting stones the way is paved for the reestablishment of reciprocity with the spirits of the low ground and the implication of *koipa maŋgiaŋ* in future fertility and prosperity.

Each adjustment in these relationships, from the hanging of the fighting stones to the rituals initiating the new cycle, requires the slaughter of pigs. It has already been pointed out that the size and rate of growth of the pig herd serves as an index to the well-being of the human population. For a local population to retain its territory after warfare obviously indicates its viability in opposition to similar local aggregates. For it to accumulate sufficient pigs to fulfill its obligations indicates its viability as an ecological and demographic unit as well.

THE DISTRIBUTION OF PORK AND THE SURVIVING PIGS

Ninety-six pigs were killed in connection with the rituals of November 7 and 8, 1963. Their total live weight is estimated to have been between 13,500 and 17,000 pounds, yielding between 6,750 and 8,500 pounds of edible meat.

---

area is, however, warranted because it may throw additional light on the structure of the ritual cycle.

What follows, it should be made clear, is my interpretation. I did not explore the possible symbolic significance of what I had observed until I had returned from the field, and it has not been possible, therefore, to learn to what extent these notions are in error, or even, perhaps, made explicit by the Tsembaga.

It may be that the entire eel ritual symbolizes the procreative act, to which it bears some detailed correspondences. The young men and boys first gathered the captured eels, which are said to be the pigs of *koipa maŋgiaŋ*, a spirit who lives in streams and who, like the water in which he lives, is explicitly associated with fertility. They carry the eels up a newly cut path through an archway where they are met by women. It is not far-fetched to regard this as sexual penetration. It is further not implausible to regard the flailing of the carcasses of the female pigs with the eels as orgasm and fertilization, with the pigs representing the female contribution to the foetus and the eels the male. Gestation is perhaps symbolized by hanging the wealth (so that it may increase) with the eels in the *timḫi* house, and by cooking the eels with the pig bellies overnight. It is also significant that the path up which the young men carry the eels *must* be a new one; this suggests defloration.

The ritual, in total, seems to me to symbolize a process of cosmic procreation, a new fertilization of both the human population and those populations of plants and animals upon which the human population most heavily depends. As such, the place of this ritual in inaugurating a *new* ritual cycle becomes apparent.

The flesh of animals dedicated to the red spirits, contributing between 1,900 and 2,400 pounds of pork to the total, was retained by the Tsembaga, as were the entrails, heads, and sometimes other parts of some of the animals dedicated to the spirits of the low ground. Exact calculation is impossible, but it may be estimated that the Tsembaga kept for themselves approximately one third, or between 2,275 and 2,635 pounds, of the pork resulting from the slaughter of their animals. This estimate of 11 to 13 pounds of pork available for every Tsembaga man, woman, and child may be lowered by about a pound per capita in consideration of the presentations made to my household. Consumption continued for five days; the meat was preserved by being suspended over fires.

The remaining 4,475 to 5,965 pounds of pork were given to members of other local populations in at least 163 separate presentations in amounts ranging from several pounds of fat or flesh to entire animals. While members of 17 other local populations were among the recipients, the Tuguma, Auŋdagai, Kauwasi, and Monambant received much the greatest part of this pork. The recipients, of course, redistributed their portions to members of their own and other groups. Almost all Tuguma, Auŋdagai, Kauwasi, and Monambant, totaling in population about 2,000, must have received some Tsembaga pig, and it is not unlikely that over 3,000 people eventually received portions from the Tsembaga slaughter.

Seventy-five animals survived the slaughter of November 7 and 8. Of these, forty were infants and juveniles, twenty-five were adolescents (weighing 96 to 120 pounds) and ten were adults (120 to 200 pounds). Of the adults, four survived sacrifice only because they had escaped to the forest and remained marked for imminent killing, two were being saved for the Tuguma *kaiko*, one belonged to a young man who was away at work and would be killed upon his return, and two were spared because, having just littered, they were thin and therefore considered to be unfit for sacrifice. Of the adolescent animals, at least four were being held for presentations to affines, with whom food taboos still prevailed at the termination of the *kaiko*. While those who had taken refuge in Kauwasi and Monambant formally abrogated the taboos on each other's foodstuffs, those who had Kauwasi or Monambant parents or grandparents were to maintain these taboos until the Monambant uprooted their *rumbim*.

The Kauwasi finished their *kaiko* between one and two years before the Tsembaga began theirs. Several of the other adolescent animals, informants said, were to be killed for the uprooting of the *rumbim* in Monambant two or three months after the Tsembaga slaughter.[14] In short, at least fifteen of the surviving pigs were scheduled for slaughter in the near future, and the surviving herd might better be regarded as comprising sixty animals averaging 60 to 75 pounds each rather than seventy-five animals.

PRESENTATION OF VALUABLES

The *konj kaiko* is the occasion for wiping the slate clean of obligations to the living as well as to the spirits, and in addition to the pork distribution, twenty-four prestations or exchanges of valuables took place at the Tsembaga *konj kaiko*.

Fourteen prestations, ranging in quantity from six to forty-two items, were made to affines for living wives or for living children that their wives had born. Two death payments were also made to affines. In one instance the brother of a woman dead over twenty years was given thirteen objects. In the other a young man made a payment for the death of his brother to his mother's clan. In addition, three exactly equivalent exchanges of valuables took place in reference to old Tsembaga women. Such exchanges between a woman's sons and her brothers or brothers' sons, end the series of payments and nonequivalent exchanges that begin shortly after marriage.

Other kinds of payments also took place. Two men who had not done so previously made prestations to men who had given them land during the exile of 1953–56, and two prestations were made in appreciation of certain services rendered. In one of these the recipient was being rewarded for carrying home from the battle ground the corpse of the donor's father. In the other a man made a small prestation to the local government-appointed head man for bringing food to his wife when a government medical patrol ordered her into the Simbai patrol post infirmary.

14. The Monambant are a Jimi Valley group. The annual cycle of plant development is somewhat different in the Jimi and Simbai Valleys, and *kaiko* in the Jimi Valley are likely to begin and end in January or February, rather than in October or November.

In one instance members of one clan "bought out" the rights of another to a young girl. Upon the death of her father, a member of the Tsembaga clan, her mother had taken up residence with her own brothers, members of another local clan, the Kamaŋgagai. Members of her deceased father's clan, wanting eventually to exchange her for a woman for one of their own boys, but recognizing that her mother's clan had rights to her because they had raised her, made a substantial payment (nine valuables) for her to her mother's brothers and their sons.

Seventeen of the twenty-four transactions were between Tsembaga and members of other local populations. I know of no instance in which a man obligated to a member of another local population did not succeed in making the payments expected of him, although in one case the amount was thought by the donor to be almost shamefully small.

It has already been suggested that the ability to stage a *kaiko* indicates the viability of the local population as an ecological and demographic unit. By fulfilling their affinal and other obligations at the *konj kaiko* through gifts of items of wealth, the members of a local population demonstrate that they comprise a viable unit in the supralocal exchange network as well.

THE KONJ KAIKO

The pig festival comes to its climax in the *konj kaiko,* when salted pig belly is publicly presented to allies and the *pave,* the ceremonial fence, is breached.

Many members of other local populations had been among the Tsembaga while the pigs were slaughtered, and during the following day many more people poured in, both in dancing formations and as individuals, from all friendly groups. By mid-afternoon the dance ground was packed with spectators and dancers; attendance may well have exceeded one thousand people by the time that the heroes' portions of salted pig belly were presented.

In the late afternoon all of the Tsembaga, with the exception of married or widowed women and infants in arms, assembled inside the *pave* enclosure and the presentation packages of salted fat were heaped behind the *pave* window. Several men climbed to the top of the structure and from there proclaimed one by one to the multitude the names and clans of the men being honored. As his name was called, each honored man charged toward the *pave* window swinging his ax and shouting. His supporters, yelling battle cries, beating drums, brandishing weapons, followed close behind him. At the window the mouth of the honored man was stuffed with cold salted belly fat by the Tsembaga whom he had come to help in the last fight and who now also passed out to him through the window a package containing additional salted belly for his followers. With the belly fat hanging from his mouth the hero now retired, his supporters close behind him, shouting, singing, beating their drums, dancing. Honored name quickly followed honored name, and groups charging toward the window sometimes became entangled with those retiring.

Between twenty-five and thirty men were publicly honored by presentations from the *pave*. Their number did not include all of those who had assisted the Tsembaga militarily in the last fight, but rather those members of other groups to whom Tsembaga were directly connected through kinship or trade and through whom support had been mobilized, as well as others who had sustained wounds and the sons of those slain. A rough order of precedence prevails, with the wounded and sons of the slain being called first. To be called last is sometimes considered an insult by the recipient, and the Tuguma who found himself in this position refused his portion. Although the Tsembaga *kaiko* was unmarred by such a conflict, it is sometimes the case that members of one local population may become angered if members of another are called before them. Vayda was witness to such a development in 1962 at the *kaiko* of the Kandambent-Namikai, at which the members of the Fungai-Korama group not only refused their pig belly but departed early.

After the presentations were completed, the visitors resumed dancing. The Tsembaga, however, still remained inside the *pave* enclosure, a section of which they now began to push down. As it fell, the men, led by those wearing *mamp gunč*, charged through the breach pounding their drums. They were followed by the unmarried girls. As they emerged the Tsembaga became a dancing formation, one of the many on the dance ground. Breaching the *pave* and subsequently dancing with contingents of friends and allies expressed, it may be suggested, breaking through the confines of debt and taboo that had long separated them from other groups.[15]

All that now remained to conclude the *kaiko* was to dig up the root of the *dawa* variety of *rumbim*, which was buried by the gate when the dance ground was built. This was quickly done, and one of the fight magic men took it and ran off the dance ground, dangling it from a string. All the Tsembaga followed him to a nearby bluff, where he disposed of it. This site overlooks enemy territory and it was in the direction of the enemy that the *rumbim* root was thrown. Then the Tsembaga removed leaves of the *rumbim* variety, *kamp*, from their buttock coverings and flung them after the exhumed root, calling out the names of enemy men and yelling, "We have finished our *kaiko*, we are here." Dancing continued through the night, and next morning, after a massive trading session, the crowd departed and the five young men, without ceremony, cut off their *mamp gunč*. The *kaiko* was finished.

THE TERMINATION OF THE TRUCE AND THE ESTABLISHMENT OF PEACE

With the termination of the *kaiko* truces that have prevailed since the last fight are also terminated, and in former times warfare usually broke out again in short order. The fighting in late 1953 or early 1954, for instance, began within three months of the end of a Tsembaga *kaiko*.

There are, however, means for reestablishing permanent peace. Peace has been made among Maring groups only very few times within the memories of even the oldest informants, and statements concerning the details of the procedure are vague and conflicting. In another publication (Rappaport, 1967) I outlined what I understood to be the practice more definitively than seems to be warranted now in light of more recent inquiries made by A. P. and Cherry Vayda. I shall first repeat my earlier account, then note the details in which it is likely to be erroneous or at least questionable.

The wreckage of the *pave* is not cleared away at the end of the *kaiko* but is allowed to rot. Its complete dissolution is said to take two to three years, about the length of time, it may be noted, that

15. In light of the previous day's eel ritual, it may have symbolized the rebirth of the local population as well.

it takes to raise a pig to full size. If fighting has not broken out by this time all adult and adolescent pigs are killed and *rumbim* and *amame* are planted after a ritual called "*pave gui*" ("black," "dead," or "rotted" *pave*). The pig herd is again allowed to reach the limits of toleration but, before the uprooting of the *rumbim*, peace-making ceremonies are jointly held by the erstwhile enemies. I shall return to these shortly.

I derived this outline, in reasonably coherent form, from the account of a single informant, a knowledgeable man, but by his own statement too young to be speaking from first-hand experience. However, it seemed to jibe with the statements of other informants, gratuitously volunteered earlier in my fieldwork, that *rumbim* would again be planted in several years (although some Tsembaga said that because of the government presence *rumbim* would never be planted again), and with the report that the Kauwasi, who had completed their *kaiko* in 1961 had *rumbim* in the ground in 1963. Because I received this information on almost the last day of my fieldwork I did not have the opportunity to investigate the matter further.

Inquiries made later by the Vaydas indicate that *pave gui* is performed to benefit the fertility and growth of pigs and requires the sacrifice of some animals, but fewer than I understood to be the case. Furthermore, they were informed that *rumbim* is not planted during or after this ceremony. Even my informant, when questioned by them, denied that *rumbim* was planted at *pave gui*. He had either been corrected between the time he spoke to me and to the Vaydas, or I misunderstood him. At any rate, it is probably the case that my earlier statement concerning this matter is incorrect.

Informants do agree, however, that peace-making rituals are put off until the pig herds of the antagonists are of large, perhaps maximal, size. This is in accordance with informants' statements to me that soon after peace is made the erstwhile antagonists attend each other's *kaiko*, and the rule I stated in the earlier publication seems generally to hold: if a pair of antagonists are able to proceed through two ritual cycles without the resumption of hostilities, they may make peace. It should be made clear, in qualification of this rule, that the rituals constituting the second cycle would be somewhat different from those constituting the first, and that the *kaiko* ending the second are, according to some informants, less elaborate. Furthermore, it is not clear that truce is assured by ritual means during the second cycle. The duration of both cycles, however, is a function of the demography of the pig herds of both parties.

Data concerning the details of the actual peace-making ritual are also deficient. Informants do agree, however, that the two sides, after each consults its own ancestors and offers them sacrifices of pig, convene at their common border where they exchange pig livers. Only those whose mothers or, in the case of long-standing enmities, those whose grandparents were members of the enemy group may eat these pig livers. The interdining taboos under which other members of the two populations labor are abrogated over four generations in accordance with the procedures described earlier.

Women are also exchanged or promised at this time. These women are explicitly regarded as "*wump*" (planting material), through which the slain can be replaced, and the children they bear are named after those whom their brothers, fathers, or grandfathers have killed. Ideally one woman should be given for each of the slain, but the Tsembaga say that this will not be possible when peace is made with the Kundagai, for the deaths have been too many. They add that many women bear more than one child, however, and it will not take many years for all the slain to be replaced by children bearing their names.

The peace-making procedure thus not only terminates the old enmity; it also, through the requirement that women be exchanged, establishes ties between former enemies. The number of new ties established by marriage is likely to be directly correlated with the severity of the former enmity, as measured by the number of deaths each of the participants has suffered at the other's hands, even when the ideal of one woman for each dead man cannot be realized.

THE RITUAL CYCLE AND AREAL INTEGRATION

The role of the ritual cycles of the Maring in interlocal relations bears upon recent discussions of areal integration in Melanesia. These discussions have focused mainly upon the ways in which the movement of goods either in trade (Schwartz, 1962) or in ceremonial exchanges (Bulmer, 1960; Salisbury, 1962) bind together groups who recognize no common superordinate authority. That the exchange system of the Maring, in which the *kaiko* is implicated, operates, like those described by other authors, to integrate all or many of the groups in an area has been implicit here. The ritual cycles of the Maring, however, provide more than a means for elaborating the relationships that arise out of economic interdependence or for formalizing in ceremonial exchanges noneconomic interdependence. When attacks may be launched, land annexed, affiliation of personnel changed, and truce or peace established are all specified in terms of the completion of events that form components of the ritual cycle. It has been shown, in turn, that the completion of these ritual events reflects the state of the relationships of the local population to both human and nonhuman components of their environment. Events in the cycle, furthermore, particularly during the *kaiko* itself, serve to assemble, bind together, and transmit information among the autonomous local populations forming the supralocal aggregates that will participate in the exercise of force. In short, the ritual cycles of the Maring provide both a means for aggregating groups of supralocal magnitude for forceful enterprises and a set of conventions governing these enterprises.

The numbers of persons whose activities are articulated through the ritual cycle of any local population are substantial. It was estimated that over 1,000 persons were present at Tsembaga for the culmination of the *kaiko,* and the groups from which dancing contingents were entertained during the course of the festival totaled over 3,000. Tsembaga pork, it was estimated, also reached 3,000 people. Simply in terms of the numbers of people involved, the aggregations formed by ritual cycles rival some of the hierarchical structures of Polynesia. The Polynesian chiefdoms may also be rivaled in the frequency with which large numbers of people are mobilized for joint activities. Information presented by Marshall Sahlins (1958:132) concerning ceremonial food redistributions in Polynesia, for instance, indicates that these events may not have occurred more often on some islands than once a year. While the *kaiko* of any local population of Maring occur with less than annual frequency, it is not unlikely that in most years a local population would be the beneficiary of a distribution arising from the *kaiko* of another group.

There are, of course, important differences between the organizations of Polynesia and the organization of the Maring. In the hierarchical, centralized organizations of Polynesia, it is reasonable to assume, system regulation was accomplished through the activities of discrete authorities, chiefs, to whom flowed information concerning system-endangering changes in the values of variables forming components of the system, and from whom flowed directives meant to return these values to safe levels. Among the Maring a consensus concerning whether or not the states of variables are in fact endangering the system must be reached, but when it is, corrective action, in accord with the conventions of the ritual cycle, follows more or less automatically. To put it in slightly different terms, the locus of the *total* ordering function in the Polynesian chiefdom is in a discrete human authority, the chief, who both *detects* deviations of variables from "safe" levels and initiates *corrective* action. Among the Maring, however, the locus of the *detecting* function is diffuse; the signals (in the form of overwork, garden damage, etc.) are received or experienced by many people and finally produce consensus that there has indeed been some system-endangering change in the state of a variable. The locus of the *correcting* function is in the ritual cycle. But in light of the fact that the activities of large numbers of people are articulated by both the highly centralized Polynesian organization and the acephalous organization of the Maring, we may ask what are the factors that account for the development or evolution of these two types of organization; in what ways might they affect differently the ecosystems in which they exist; what precisely are the differences in their capacities for articulating the activities of large numbers of people; where lie their comparative strengths and weaknesses. Similar questions have been posed recently by Sahlins (1963) and Schwartz (1962). These questions require further research. It may be suggested, however, that the Polynesian form of system regulation is considerably more sensitive than the Maring.

First, the strength of the signal required to initiate corrective action in the chiefdom can be considerably weaker than is likely to be required in the Maring system, for only one man, and not the number of people sufficient to form a consensus, need detect system-endangering trends in the states of variables. Second, a variety of corrective programs, differing from each other both formally and in magnitude, may be initiated by the chief, in contrast to the stereotyped ritual response of the Maring. In short, Polynesian system regulation operates more rapidly and more flexibly than that of the Maring. This might permit the tighter and more continuous coordination of the activities of larger numbers of people than is possible among the Maring. It might also be argued that more sensitive system regulation, such as that found in Polynesia, might avoid the wide fluctuations in the value of variables noted in the Maring area (e.g., in the size of pig herds and the amounts of land in cultivation). It must be emphasized, however, that in the Maring case the fluctuations themselves are of great value, for they are critically important in the regulation of intergroup relations.

Maring system regulation also has its advantages. No special personnel, such as chiefs and their retainers, with elaborate requirements (sumptuary goods, elaborate assembly structures, and the like) need be supported. Furthermore, the corrective responses set by the conventions of the ritual cycle, although they are slow and stereotyped, have a certain merit, given a more or less stable cultural environment: they provide little room for human error. It may also be suggested that it is their very inflexibility that makes it possible for ritual cycles to regulate relationships between autonomous local groups, particularly in respect to warfare.

CHAPTER 6

Ritual and the Regulation of Ecological Systems

The place of ritual in the ecology of the Tsembaga has been the focus of this study. In the earlier chapters the material relations of the Tsembaga with their environment were described, and in the preceding chapter the regulatory functions of ritual were discussed. The Tsembaga ritual cycle has been regarded as a complex homeostatic mechanism, operating to maintain the values of a number of variables within "goal ranges" (ranges of values that permit the perpetuation of a system, as constituted, through indefinite periods of time). It has been argued that the regulatory function of ritual among the Tsembaga and other Maring helps to maintain an undegraded environment, limits fighting to frequencies that do not endanger the existence of the regional population, adjusts man-land ratios, facilitates trade, distributes local surpluses of pig in the form of pork throughout the regional population, and assures people of high-quality protein when they most need it.

Despite the importance attached to religious ritual, which is, so far as we know, a specifically human phenomenon, the frame of reference within which it and other components of culture have been viewed in this study has been borrowed from animal ecology. The Tsembaga, designated a "local population," have been regarded as a population in the animal ecologist's sense: a unit composed of an aggregate of organisms having in common certain distinctive means whereby they maintain a set of shared trophic relations with other living and nonliving components of the biotic community in which they exist together.

Tsembaga territory, moreover, has been regarded as an ecosystem, a demarcated portion of the biosphere that includes living organisms and nonliving substances interacting to produce a systemic exchange of materials among the living components and between the living components and the nonliving substances. This demarcation was guided by particular analytic goals but it is not completely arbitrary. Ecosystems are defined in terms of trophic exchange, and the Tsembaga alone among humans are directly involved in trophic exchanges with the nonhuman entities with which they share their territory. Conversely, the Tsembaga are not directly involved in trophic exchanges with nonhuman entities in the territories of other local populations. This demarcation of ecosystem boundaries is not unduly anthropocentric, for the cycle of materials in which trophic exchanges result is generally highly localized in tropical rain forests.

Although it is possible, through the application of criteria borrowed from animal ecology, to designate the territory of the local group as an ecosystem, it must not be forgotten that the environment of any local human group is likely to include more than those entities in its immediate locality upon which it subsists and which subsist upon it. Other components of the external world affect its survival and well-being in other ways. Neighboring human groups are hardly less significant to Tsembaga survival than the secondary forest in which they plant their gardens and are perhaps more significant than the primary forest in which they trap marsupials. As they participate in a set of trophic exchanges with the members of other species with which they share their territory, so do they exchange genetic materials, personnel, and goods with members of other local populations occupying other territories. Furthermore, it is within this larger field that land is redistributed through warfare. These supralocal relations, it seems to me, can hardly be ignored in ecological analyses, and I have given them as much attention in this study as the more localized trophic relations.

The concept of the ecosystem, however, which provides a convenient frame or model for the analysis of trophic exchanges between ecologically dissimilar populations occupying single localities can accommodate only by the introduction of analogy nontrophic material exchanges between ecologically similar populations occupying separate localities. Instead of extending the concept of the ecosystem (a system of localized trophic exchanges) to include (typically nontrophic) exchanges between distinct local populations of humans exploiting separate areas, I suggest that we recognize that local populations of humans (and many other species as well) are also likely to participate in regional systems. These systems, as I have already indicated, will include among their most important components the several local human populations that occupy distinct areas within general regions. Such aggregates of local populations may be called "regional populations."

It is worth noting that regional populations—aggregates distinguished by the criteria of regional continuity and exchanges of personnel, genetic material, and goods—may in some instances be more or less coterminous with other aggregates distinguished by anthropologists, ethologists, and geneticists by application of other criteria. These are "societies," aggregates of organisms that interact according to common sets of conventions; and "breeding populations," aggregates of interbreeding organisms capable of persisting through an indefinite number of generations in isolation from similar aggregates of the same species. Societies, breeding populations, and regional populations are similar in that they are likely to persist and evolve through long periods of time while their constituent subunits (such as the Tsembaga) are relatively ephemeral. It may be, therefore, that in long-range evolutionary studies these more inclusive aggregates, rather than one or another of their constituent subunits, should be central to the analysis.

It is as difficult to establish boundaries for the regional population as it is for the society or breeding population by any nonarbitrary means. Therefore we can only say that these differently defined aggregates may, in some instances, be roughly coterminous. The notion of clines may be as useful in social and ecological anthropology as in genetic inquiry, but in a study such as this we are concerned only with the ways in which a single local population relates to others of its kind during a particular period in its history, and precise definition of the extent of the regional population is not necessary. It is sufficient to note that the regional population of which the Tsembaga are a constituent unit includes the Maring speakers and their neighbors.

The analytic strategy I have followed in this study involves, then, the discrimination of two systems, the ecosystem and the regional system, with the local population participating in both.[1] I believe that this procedure has certain advantages. First, it protects the power of the ecosystem concept by preserving it from the analogies that would inevitably become necessary if it were extended to include all external relations of human groups. Second, it enhances the utility of the ecosystem concept by enabling us to designate as ecological (local) populations and ecosystems units of sufficiently small size to permit convenient quantitative analysis. These can often be "natural units," units with an existence independent of our discrimination, such as recognized social groups and their land holdings. In this study, named local communities were taken to be populations and the areas they exploit in subsistence activities as ecosystems. Because of poor group or territorial definition, economic complexity, or other reasons it may not always be possible to find such convenient natural units, but where it is possible the advantages may transcend mere convenience. It has already been mentioned, for instance, that the Tsembaga and coordinate groups, being territorial, are in fact the most inclusive human groups directly involved in trophic exchanges within the Maring area. Therefore, their identification as local populations has descriptive validity as well as analytic utility, for it reflects the manner in which the Maring and their neighbors are distributed in space and in relation to their resources.

The discrimination of the regional system and the ecosystem also aids us in illuminating the systemic functions of certain cultural phenomena. I have distinguished the regional system from the ecosystem by differences in the material exchanges that typify them. These differences imply a more abstract criterion by which the two systems may be distinguished: the criterion of internal coherence.

Fully coherent systems are those in which a change in the state of any single component immediately results in proportional

1. I believe that the discrimination of two (sub) systems is sufficient to accommodate the relatively simple Maring ecology, but I recognize that other, more complex, situations might well require more elaborate discriminations.

For example, in the case of a coral atoll, it might be analytically useful to designate that portion of the biosphere within which the human population is directly involved in trophic exchange as the "immediate environment," and to regard it to include three more or less distinct ecosystems, the terrestrial, the reef-lagoon, and the open sea. Such a procedure would recognize both the important differences in the biotic communities in the three areas, and differences in man's participation in them.

Similarly, it may sometimes be necessary or useful to regard the "nonimmediate" environment to be composed of several regional exchange systems.

changes in the states of all other components. At the opposite extreme are completely incoherent systems, for which the term "heap" is sometimes used; the state of any entity in the heap may vary without affecting the states of any of the others. It is obvious that no system in which organisms participate which is not trivial to their survival can be completely incoherent. Conversely, too much coherence may also endanger the survival of participating organisms.

So far as I know no measures of coherence have been devised; nevertheless we may discern in nature collections of entities that affect each other more markedly and more rapidly than they affect or are affected by others. Individual organisms, for instance, show great internal coherence, but less coherence with respect to entities external to themselves. I believe that a high degree of internal coherence also characterizes the ecosystem on the one hand and the regional system on the other, and that it is therefore legitimate to distinguish them from each other.

Nevertheless, the systemic discontinuities that enable us to distinguish sets of highly coherent phenomena from each other (and designate them as systems) are only relative. Nature is continuous, and the several systems one may establish are not likely to be independent of each other. For one thing, they are likely to share some components, and, through these, events in one system eventually are likely to affect events in others. In this study events in the separate ecosystems of the local populations eventually affect the regional system, and vice versa. Since this is the case, the systems we discriminate by application of material criteria, such as types of exchange, or abstract criteria, such as coherence, are properly regarded to be *subsystems* of larger systems that together they comprise.

I would suggest that the strategy of distinguishing subsystems has more general applicability than the ecological use to which it has been put here. It has been useful in this study to distinguish two subsystems, but doubtless several could be distinguished among the phenomena considered in any ethnographic study, and important questions would follow. How are the various subsystems in which any human aggregate participates articulated? That is, through what mechanisms do changes in one subsystem effect changes in others? What items in the culture are involved? Are the effects of changes in one subsystem reflected in others continuously, periodically, or only when some threshold value is transgressed? In the subsystems that are involved, are the changes proportional? "Classical" functionalism, with its assumption of the interrelatedness of phenomena, generally has been concerned with questions similar to these. It can nevertheless be asserted that the search for mechanisms that articulate subsystems and the elucidation of how such mechanisms work can be made more strictly operational by specifying as precisely as possible the subsystems that are in fact being articulated.

It has been argued in this study that Maring ritual is of great importance in articulating the local and regional subsystems. The timing of the ritual cycle is largely dependent upon changes in the states of components of the local ecosystem. But the *kaiko,* which culminates the ritual cycle, does more than reverse changes that have taken place in this subsystem. It also affects relations among the components of the regional subsystem. During its performance obligations to other local populations are fulfilled, support for future military enterprises is rallied, land from which enemies have earlier been driven is occupied, and the movement of goods and women is stimulated. Completion of the *kaiko* permits the local population to initiate warfare again. Conversely, warfare is terminated by the *rumbim*-planting ritual that prohibits the reinitiation of warfare until the state of the local ecosystem allows the *kaiko* to again be staged and completed; participation in the *rumbim* planting also ratifies the connection of men to local populations to which they were not previously affiliated.

Maring ritual, in short, operates not only as a homeostat—maintaining a number of variables that comprise the total system within ranges of viability—but also as a transducer—"translating" changes in the state of one subsystem into information and energy that can produce changes in the second subsystem. It should be recalled here that the transduction operation of the ritual cycle is such that the participation of local populations in respect to warfare, which is important in the redistribution of land and personnel but is also dangerous, is not continuous. It could therefore be argued that the ritual transducer maintains coherence between subsystems at levels above or below which the perpetuation of the total system might be endangered.

In the functional and cybernetic analysis of the ecological relations of the Tsembaga that has been attempted here, variables have been abstracted from events or entities in the physical world and treated as components of an analytic system. If the method of functional analysis is to meet the criticisms of Hemple (1959) and others, quantitative values must be assigned to all variables, and in the case of those variables that define the adequate functioning of the system their tolerable ranges of values must also be specified.

This study, however, has left some variables unquantified. No values were assigned, for instance, to the frequency of warfare, much less to its tolerable limit. While upper limits were assigned to the tolerable range of size of the human and pig populations, lower limits were not. It was suggested that a population's tolerance for the destructiveness of pigs is limited, but this limit was not assigned any quantitative value. Since these and other quantitative data are lacking, this study remains in part what Collins (1965) terms an "explanatory sketch."

Appropriate cautions about the reliability of values assigned to variables were issued or were implicit in the course of discussion as the values were assigned, and the methods by which the values were derived have been spelled out in the appendices. It bears repeating, however, that the measuring techniques were sometimes crude and in some instances statistically insufficient. It should be kept in mind that these data result from the efforts of an anthropologist without special training in such fields as botany, nutrition, physiology, and surveying. To point this out is neither to defend nor to denigrate the worth of the data. It is simply to characterize them and to draw attention to the need, if ecological studies are to show increased refinement, for anthropologists either to broaden their training or to be supported by personnel with training in other fields.

The systemic role of ritual in the ecology of the Tsembaga has been the focus of this study and I have offered no suggestions concerning the origin of these rituals or any of the other phenomena described here. As Durkheim pointed out in 1895:

> To show how a fact is useful is not to explain how it originated or why it is what it is. The uses which it serves presuppose the specific properties characterizing it but do not

> create them. The need we have of things cannot give them existence, nor can it confer their specific nature upon them. It is to causes of another sort that they owe their existence. [1938:90]

It may be mentioned, however, that ritual components that are formally similar to those of the Tsembaga are found among other New Guinea highland groups. Massive pig slaughters are, of course, widespread, and such elements as "red spirits," ceremonial fences, and structures resembling *timbi* houses have been reported for a number of groups (Bulmer, 1965; Luzbetak, 1954; Read, 1955; Reay, 1959; Newman, 1964). The ritual use of *Cordyline fruticosa* is also extremely widespread, being found as far away as the Philippines (H. C. Conklin, personal communication). The way these components are arranged in particular events and in particular ritual cycles, however, seems to show considerable variation. The data are not sufficient, but it may be suggested, in light of the functions of ritual among the Tsembaga, that ritual regulation may be widespread in New Guinea, and that variations in ritual may be the result of changes brought about by differences in the ecological circumstances of various populations. Random cultural changes, analogous to gene drift, may of course be involved, but it is not implausible to suggest that as variables and the relations among them change so do the mechanisms that regulate them.

Another point should be made explicit here. This study has been concerned with aspects of Tsembaga and Maring environment, physiology, demography, psychology, the economic, social, and political structure, and religion: phenomena falling into classes that have frequently been assigned to several ontological "levels" (inorganic, organic, superorganic). Some social scientists have argued strongly that events or processes occurring in each level are essentially autonomous in respect to events and processes occurring on other levels, and that explanations that cut across levels are either reductionistic or the opposite.

The notion of levels has been a useful device for organizing science, dividing the study of nature, so to speak, among the various disciplines. The sociologist O. D. Duncan has recently recognized the importance of the concept of levels in the sociology of science, stating, somewhat pejoratively, that "its major contribution to the history of ideas has been to confer legitimacy upon the newer scientific approaches to the empirical world that, when they were emerging, had good use for any kind of ideological support" (1961: 141).

Duncan is perhaps too harsh. In addition to its usefulness in the organization of science, the concept of levels may also be of use in the ordering of data, particularly when explanations of form are sought. But at best it is only one way of ordering data. The concept of the system, which presupposes the interdependent variation of a number of elements, is another. Those who would argue for the functional autonomy of levels apparently assume that organic phenomena, for example, as a general class show greater internal coherence than do any of its members (e.g. organisms, populations) with superorganic, or cultural, phenomena. It seems to me that such a position confuses ontological status with systemic interrelations—confuses what things are "made out of" with causes and effects of their behavior. Systems discovered in or abstracted from relationships observed in nature may, and frequently do, cut across ontologically defined levels.

"Manifest" and "latent" functions have been suggested for many of the rituals discussed in this study. Some of these proposals, particularly those concerned with the role of ritual and taboo in the containment of conflict, have been in the nature of post facto interpretations of observed or reported events and are probably difficult to validate empirically. As such, they are vulnerable to the criticisms leveled against "classical" functional analysis by such writers as Carl Hempel (1959). But these suggestions have been only incidental to the main concern of this study. The emphasis here has been upon the homeostatic function of ritual, and upon the ways in which it links subsystems.

Religion as a homeostatic mechanism has been discussed by Miller (1964), who argues against a growing tendency to discriminate in human phenomena separate ontologically defined "systems" (e.g. cultural system vs. social system, religious system vs. secular system) and states that "cultures are the viable systems for study, and are the *only* wholes we can isolate. Cultures have a 'command pattern,' a structure which is extended throughout the system . . . and keeps it viable as a system" (p. 94, his italics). "Command" is earlier defined as "the locus of the ordering operation," and is identified by Miller with religion.

That ritual and the understandings that elicit ritual behavior comprise the "locus of the ordering operation" in the systemic relationships with which this study has been concerned does not need further reiteration. The examination of the role of ritual among the Tsembaga would suggest disagreement only with Miller's assertion that "cultures are the viable systems for study, and are the *only* wholes we can isolate." Culture has been regarded here not as itself a whole, but as a part of the distinctive means by which a local population maintains itself in an ecosystem and by which a regional population maintains and coordinates its groups and distributes them over the available land. That which is regulated by the command pattern of belief and ritual does not consist merely of the interrelations of other components of the culture, but also includes biological interactions among organisms not all of which are human. Indeed, it would not be improper to refer to the Tsembaga and the other entities with which they share their territory as a "ritually regulated ecosystem," and to the Tsembaga and their human neighbors as a "ritually regulated population."

In light of the analysis presented in this study it may be asked whether rituals have peculiar virtues that make them particularly well suited to function as homeostats and transducers. I can offer only brief and highly speculative suggestions here.

It is worth noting, first, that in mechanical, electronic, and physiological systems, in which the states of other components may vary through a continuous range, the range of states of the regulating mechanism is frequently limited to two. A thermostat, for instance, is basically a switch that, in response to a particular amount of change in the medium in which it is immersed, goes "on" or "off," thus activating or deactivating a source of heat. If we were to represent a heating system in the abstract we would regard the thermostat as a two-valued, or binary, variable, while the other components of the system would be treated as continuous, or many-valued, variables. One of the great advantages of the binary regulating device is simplicity. Indeed, its response to the continuously changing states of other components of the system is the simplest conceivable:

if these changes exceed certain limits, the binary mechanism switches from one to the other of its two possible values.

It may be that the very simplicity of the binary mechanism minimizes the likelihood of its breakdown, but a more important aspect of binary operation has already been suggested in another context. Binary control eliminates the possibility of error from one phase of system operation: an inappropriate response cannot be selected from a set of possible responses because the set of possible responses has only one member. To put it in anthropomorphic terms, the regulating mechanism, once it receives a signal that a variable has transgressed its tolerable range, does not have to decide what to do. It can do only one thing or nothing at all.

Like thermostats, rituals have a binary aspect. As the thermostat switches on and off, affecting the amount of heat produced by the furnace and the temperature of the medium, so the rituals of the Tsembaga are initiated and completed, affecting the size of the pig population, the amount of land under cultivation, the amount of labor expended, the frequency of warfare, and other components of the system. The programs that should be undertaken to correct the deviation of variables from their acceptable ranges are fixed. All that need be decided is whether in fact deviations have occurred. The Tsembaga reach such decisions through discussion and then the formation of a consensus.

It must be recognized that while Maring ritual regulation enjoys the advantages of simplicity, it also suffers from simplicity's limitations. Consensus regarding a deviation from acceptable conditions is likely to be slow to form and the programs initiated to correct such changes are inflexible and unlikely to be proportional to the deviation. In a stable environment slow and inflexible regulation may not produce serious problems, but the novel circumstances that are continuously presented by rapidly changing environments may require more rapid and flexible regulation.

The binary aspect of rituals is also important with respect to their role as transducers. The mere occurrence of a ritual can be regarded as a signal. Since a ritual can, at a particular time, only be occurring or not occurring, its occurrence transmits binary information.[2] Binary

2. It was only when this study was already in proof that I read Anthony F. C. Wallace's *Religion: An Anthropological View*. Wallace states (p. 233) that "ritual

information is qualitative in that it is information of a "yes-no," rather than a "more-less," sort. The performance of a ritual, however, may depend, as many Maring rituals do, upon a complex set of quantitative relationships among many variables. Thus, *the occurrence of the ritual may be a simple qualitative representation of complex quantitative information.*

The importance of this aspect of ritual function may be illustrated by reference to the uprooting of *rumbim*. This ritual can be regarded as a statement concerning the complex of quantitative conditions prevailing in a local subsystem (ecosystem) at a particular time. Now the quantitative information that the qualitative ritual statement (uprooting the *rumbim*) summarizes is not available to populations other than the one performing the ritual, and even if it were it would be subject to perhaps erroneous interpretation. Being summarized are not merely the constantly fluctuating values of a number of separate variables but the continually changing relationships among these variables. It would be difficult indeed to translate quantitative information concerning the constantly fluctuating state of the local subsystem directly into terms that would be meaningful to other populations in the regional subsystem. The information would be ambiguous at best. But this difficulty is overcome if a mechanism is available to summarize the quantitative information and translate it into a qualitative signal.

Uprooting the *rumbim* is such a mechanism. The virtue of this ritual is that it signals *unambiguously* that the local subsystem has achieved a certain state, and that, therefore, the local population may now undertake previously proscribed actions that are likely to affect the regional subsystem. The absence of ambiguity from this message derives from the binary character of the ritual trans-

---

may . . . be classified as communication without information: . . . each ritual is a particular sequence of signals which, once announced, allows no uncertainty, no choice, and hence, in the statistical sense of information theory, conveys no information from sender to receiver." My discussion here does not refer to the sequence of signals *within* a ritual. It takes the *occurrence* of the ritual to be a signal. I am arguing that the occurrence of a ritual, even one that occupies a particular position in a fixed sequence of rituals, does convey information if (1) the time of its occurrence is not calendrically fixed or (2) if the time of its occurrence is not separated by a standard interval from the preceding or subsequent ritual in the sequence. Despite this possible disagreement the remainder of this discussion, as well as preceding ones (particularly that concerning *Kaiko* as display), would have benefited from a consideration of Wallace's enlightening study.

duction device, which reduces a great complex mass of "more-less" information to a simple "yes-no" statement. It is interesting to note here that control transduction in physiological systems may rely heavily upon binary mechanisms and information (Goldman, 1958: 116ff).

Binary mechanisms thus make suitable regulators and transducers. But why is it that such mechanisms should be embedded in religious practice? In other words, what advantages does sanctity confer upon transducers and homeostats? As far as transduction is concerned, I can only elaborate here a suggestion made earlier in the discussion of ritual as display. A ritual, such as uprooting the *rumbim,* has only a conventional relationship to the ecosystemic state that it is taken to represent. As such, it can be regarded as a symbol. Any form of communication that employs symbols can accommodate lies. But a ritual is not only an act of communication; it is also a sacred performance. Although sanctity inheres ultimately in conceptions that are not only assumed by the faithful to be true but whose truth is placed beyond question or criticism, objects and activities associated with these conceptions partake of their sanctity. Since that which is sacred is taken by the faithful to be unquestionably true, sanctified messages are more likely than unsanctified messages to be accepted as true. Conversely, the transmitter's fear of supernatural sanctions may induce him to refrain from deliberately sanctifying and transmitting false information. Sanctification, in short, may enhance the reliability of symbolically communicated information. (For a similar discussion see Waddington, 1960.)

Another aspect of sanctity is important with respect to homeostats. It may be noted that the Maring are without powerful authorities, authorities which have at their disposal men and resources that can be organized to exert force upon the physical and social environment.[3] Indeed, if authorities are defined as discrete loci in communications networks from which directives emanate, it is difficult to identify authorities among the Maring at all. In the near absence of authorities the conventions of the ritual cycle specify the courses of action (such as sacrificing pigs) or inaction (the observance of truces) to be undertaken at specified times or during specified

3. Following Bierstadt's suggestion (1950:737), I regard political power to be the product, in a mathematical sense, of men, resources, and organization.

periods. In the absence of power vested in discrete authorities, compliance with the conventions is ensured, or at least encouraged, by their sanctity. Sanctity, thus, is a functional alternative to political power among the Maring, and no doubt among other people as well. Among the world's peoples we perhaps could discern a continuum from societies, such as the Maring, that are governed by sacred conventions in the absence or near absence of human authorities through societies in which highly sacred authorities have little power to societies in which authorities have little sanctity but great power. It would be plausible to expect this continuum to correlate roughly with technological development, for technological sophistication is likely to place highly effective weaponries in the hands of authorities, weaponries not generally available to their subjects. An authority with great power can dispense with sanctity—as Napoleon said, "God is on the side of the heavy artillery."

Although this study has been primarily concerned with the role ritual plays in the material relations of the Tsembaga, it is nevertheless the case that the Tsembaga say that they perform their rituals to rearrange their relations with spirits. It would be possible in an analysis of the empirical consequences of ritual acts to ignore such rationalizations, but anthropology is concerned with elucidating causes, as well as consequences, of behavior, and proximate causes are often to be found in the understandings of the actors. It seems to me, therefore, that in ecological studies of human groups we must take these understandings into account.

I have suggested elsewhere (Rappaport, 1963:159; 1967:22) that two models of the environment are significant in ecological studies, and I have termed these the "operational" and the "cognized." The operational model is that which the anthropologist constructs through observation and measurement of empirical entities, events, and material relationships. He takes this model to represent, for analytic purposes, the physical world of the group he is studying. The Tsembaga environment has been represented as a complex system of material relationships composed of two subsystems distinguished from each other by differences in the materials exchanged in each, but affecting each other through mechanisms amenable to direct observation.

The cognized model is the model of the environment conceived by the people who act in it. The two models are overlapping, but not identical. While many components of the physical world will be represented in both, the operational model is likely to include material elements, such as disease germs and nitrogen-fixing bacteria, that affect the actors but of which they may not be aware. Conversely, the cognized model may include elements that cannot be shown by empirical means to exist, such as spirits and other supernatural beings.

Some elements peculiar to the cognized model may be isomorphic with elements peculiar to the operational model. The Tsembaga say, for instance, that they are loath to build houses below 3,500 feet because certain spirits that are abroad at night in low areas give one fever. The behavior of these spirits—and the consequences of their behavior—corresponds closely to that of anopheles mosquitoes, which the Tsembaga do not recognize to be carriers of malaria. But elements, and relationships among elements, in the two models need not always be isomorphic or identical. The cognized and operational models may differ in some aspects of their structure as well as in the elements included in each.

This is not to say, of course, that the cognized model is merely a less adequate representation of reality than the operational model. The operational model is an observer's description of selected aspects of the material world. It has a purpose only for the anthropologist. As far as the actors are concerned it has no function. Indeed, it does not exist. The cognized model, while it must be understood by those who entertain it to be a representation of the material and nonmaterial world, has a function for the actors: it guides their action. Since this is the case, we are particularly concerned to discover what the people under study believe to be the functional relationships[4] among the entities that they think are part of their

4. The difficulties of attempting to discover the native, or cognized, model of the environment are even more severe than the difficulties of trying to construct an operational model. The method of ethnoscience pioneered by Conklin (1957), Frake (1962), and others, while valuable, has largely been directed toward the elucidation of native taxonomic distinctions, and taxonomic distinctions do not necessarily indicate folk notions of functional processes (Vayda and Rappaport, in press). The exploration of native ideas about functional relationships still rests, it seems to me, on the rather impressionistic methods that have long prevailed in anthropology, although it may be that a structural approach similar to that advocated by Claude Levi-Strauss

environment, and what they take to be "signs," indicating changes in these entities or relationships, which demand action on their part; but *the important question concerning the cognized model, since it serves as a guide to action, is not the extent to which it conforms to "reality" (i.e. is identical with or isomorphic with the operational model), but the extent to which it elicits behavior that is appropriate to the material situation of the actors, and it is against this functional and adaptive criterion that we may assess it*. Maring notions of disease etiology are certainly inaccurate, but the slaughter and consumption of pigs during illness is just as effective when undertaken to strengthen or mollify spirits as it would be if it were specifically undertaken to alleviate stress symptoms.

It may even be that lack of correspondence between some aspects of the cognized model and the real world which it is taken to represent confers a positive advantage upon a population (Vayda and Rappaport, in press). For instance, it might be to the immediate material advantage of a large local Maring population to attack a weaker neighbor in violation of the ritual truce or without signaling the possibility of such an attack by uprooting the *rumbim* and staging a *kaiko*. Although there have been exceptions, Maring populations are not likely to act in this fashion because their members believe it would be to their disadvantage. They would fear that their ancestors would not support them and that their enterprise would therefore fail. It has been asserted that ritual truces are advantageous to the Maring as a whole because through them the occurrence of warfare is limited to tolerable frequencies. This advantage is achieved by masking from some constituent local populations an appreciation of where their own immediate material interests lie. It can thus be argued that the cognized model is not only not likely to conform in all respects to the real world (the operational model) but that it must not.

The cognized model of the environment, then, is understood by the functional anthropologist to be part of a population's means of adjusting to its environment. It may be suggested that the place of the cognized model in the material relations of a population is analogous to that of the "memory" of a computer control in an

---

and Edmund Leach or the "cultural grammars" suggested by some other writers could be of use.

automated system of material exchanges and transformations. In the automated system, signals concerning the states of variables are received in the memory, where they are compared to "reference" values or ranges of values. (These values, although they are stored in the memory, may themselves be continuously adjusted in response to signals from other parts of the system.) In response to discrepancies between actual and reference values, programs are initiated that tend to return the deviant variable toward states approximating reference values or within reference ranges. Ideally, a corrective program is discontinued when the discrepancy between the signal emanating from a system component and the reference value is eliminated. Powers, Clark, and McFarland (1960) argue that similar information feedback governs the behavior of individuals, and I have suggested here that it is important in group behavior as well.

It is reasonable to assume that people compare the states of components of their environment, *as these states are indicated by signs,* with their notions (reference values or ranges) of what these states should be. For example, I have stated that a ritual cycle culminates when women's complaints and garden invasions by pigs, both signs that the pig herd is becoming burdensome, transgress the limits of a reference range. No value could be assigned this limit; it was only defined as complaints and invasions of significant number, magnitude, and frequency to affect enough people to shape a consensus. The transgression of this limit is itself a sign that there are sufficient pigs to repay allies and ancestors, and a *kaiko,* which is, among other things, a program directed toward returning the values of pig-related variables to the reference range, is initiated. With the completion of the *kaiko,* garden invasions by pigs are reduced to zero, or nearly so, and the women are left to complain about other things. The discrepancy between signals (complaints of overwork and damage) emanating from system components (affected members of the group) and the reference range has been abolished.

Some important questions follow. First, what is the relationship between signs and the processes they are taken to indicate? Is it the case, for instance, that such processes as environmental degradation are detected (indicated by signs) early, or only when they are well advanced? Second, to what extent do reference values, which are likely to reflect people's wants rather than their needs, correspond to the actual material requirements of the local population, the ecosystem, or the regional population? In other words, *what is the relationship between the reference values or ranges of values of the cognized model and the goal ranges of the operational model?* Concerning the Tsembaga I have noted, for instance, that the upper limit of the reference range for pigs (as defined by complaints) is likely to be below the limit of the goal range, as defined by the carrying capacity of the territory.

This study has been concerned with regulation, or processes by which systems maintain their structure, rather than adaptation, or processes by which the structure of systems change in response to environmental pressures. I shall therefore only suggest here that if cognized models are important components of control mechanisms their consideration in evolutionary as well as functional studies is warranted. We may ask in what ways cognized models change in response to environmental pressures, what differences there may be in the abilities of various cognized models to modify themselves, and how change, or resistance to change, in the cognized model affects the material relations of a people. In other words, native epistemology may be of considerable importance in evolutionary processes. Lawrence, for example, has recently argued (1964), in effect, that the inflexibility of the cognized models of various peoples of the Madang District has seriously hampered their efforts to adapt to material changes in their environment following contact with Europeans.

I have hardly touched upon change in cognized models here (one relevant suggestion was made in passing in footnote 9, Chapter 4), and I have not treated the relationship between Maring cognized and operational models adequately. Nevertheless, it is worth raising such questions, even at the end of this study, for through them we may, in future studies, be able to inquire, in reasonably well-defined ways, into the functional and adaptive characteristics of ideology.

My belief that it is not only possible but preferable to examine man's ecological relations in terms that also apply to noncultural species has been reflected throughout this study. Anthropology has been mainly concerned with phenomena unique in man, but it seems to me that if we are to understand what is uniquely human we must also consider those aspects of existence which man shares with other creatures. This conviction has led me to set religious rituals and the beliefs associated with them in a frame of reference that can also accommodate the behavior of animals other than man. It is this frame of reference that has exposed the crucial role of religion in the Marings' adjustment to their environment. The study of man the culture-bearer cannot be separated from the study of man as a species among species.

APPENDIX 1

Rainfall

Table 11. Rainfall in Tabibuga and Dikai (Tsembaga Territory)

	Tabibuga				Dikai		
Month	Lowest monthly rainfall 1959–1963 in points*	Highest monthly rainfall 1959–1963 in points	Average monthly rainfall 1959–1963 in points	Monthly rainfall Dec. 1, 1962–Nov. 30, 1963 in points	Monthly rainfall Dec. 1, 1962–Nov. 30, 1963 in points	Number of days rain fell	Number of days sun shone
December	664	1,626	1,226	1,626	1,879	29	18
January	550	2,167	1,216	586	770	22	27
February	365	2,001	1,386	365	676	14	28
March	1,420	1,823	1,511	1,432	1,106	incomplete	incomplete
April	568	2,044	1,437	568	1,171	26	20
May	391	1,410	832	391	873	18	20
June	199	799	507	779	900	24	21
July	260	1,211	589	410	703	18	24
August	338	873	637	733	1,495	28	24
September	610	1,401	897	1,401	2,020	26	20
October	676	1,769	1,107	1,769	1,919	27	23
November	974	1,734	1,086	943	1,877	21	18
	Lowest yearly total	Highest yearly total	Average annual total	Annual total	Annual total		
Totals in points	11,003	14,068	12,888	11,003	15,389	253+	243+

* 1 inch = 100 points.

APPENDIX 2

Soils

Data on soils are scanty. One hundred eleven samples were taken from various locations on Tsembaga territory. Unfortunately these were delayed in transit for almost one year, during which time they were damaged; only nine could be salvaged. The results of their analysis by Dr. Hugh Popenoe, director, Department of Soils, University of Florida, are presented in Table 12. Dr. Popenoe's discussion of these results is summarized here.

The fact that the pH values of the soils are low, although the nutrient contents are high in comparison to them, would indicate that the samples are probably very high in organic materials. In most tropical forest soils high organic matter contents are found in the top soil, in the surface litter, and A horizons, below which the organic material diminishes and the mineral content increases. Samples 2, 3, and 4 constitute examples of this point. Although sample 3, from the A horizon (*gi miña:* black earth), which terminates at a depth of only six inches, has a low pH, its nutrient content is much higher than that of samples 2 and 4, which were taken from the B horizon (*miña añeŋgi:* red earth) at the same site (sample 2 was taken from the same boring).

The nutrient contents of all three of these samples is low, although that of sample 3 is adequate for crop production. Since, however, this is the top soil, and fertility is mostly tied up in the organic fraction, these nutrients would, according to Dr. Popenoe, probably be lost in a year or two of heavy cropping. (Street and Clarke question this point.) The surface litter from this location was among the samples lost. In this location it was 1 to 1.5 inches in depth, and if it were of a richness comparable to either sample 42 or 69 it would enhance fertility, at least in the short run, considerably. The site from which these samples were taken was a newly cut but not yet burned garden in secondary forest estimated to be twenty to twenty-five years old.

Table 12. Tsembaga Soils

Sample number	Swidden number	Year planted	Altitude	Ground name	Vegetation when sample taken	Depth	Horizon	Native soil type	pH	CaO ppm	MgO ppm	P_2O_5	K_2O	NO_3	O.M.
2	68B	1963	4,400	Tipema	advanced second forest (not yet cut)	Below 6″	B	añeŋgi (red)	5.2	619	218	Trace	24	very low	
3	68B	1963	4,400	Tipema	advanced second forest (not yet cut)	1″–4″	A	gi miña (black)	4.5	1,029	880	0.5	501	very high	
4	68B	1963	4,400	Tipema	advanced second forest (not yet cut)	4″–7″	B	?	4.4	208	382	1.0	232	very high	
13	Oa	1960	4,400	Gomrup	young second-growth trees and grasses	1″–4.5″	A	gi	4.8	833	428	1.0	459	very high	
42	25A	1961	4,800	Yindokai	sweet potatoes, secondary growth	0–1.2″	litter surface	ñeñuŋ	5.1	8,928	3,676	14.0	666	very high	high
67	—	1958	4,200	Gerki	Imperata; few trees	1.2″–3.4″	A	gi	4.6	645	382	1.0	146	very high	
69	19A	1962–63	5,000	Tendopeŋ	crops	0–1.2″	litter surface	ñeñuŋ	5.2	7,526	702	6.0	542	very high	high
74	—	1959	4,400	Gomrup	Imperata; few trees,	Below 10″	B	añeŋgi	5.0	99	498	Trace	73	very low	
89	—	1943	3,500	Porakump	advanced secondary forest	Below 8.1″	B	añeŋgi	5.3	4,995	1,562	2.5	792	very high	high

Sample 13 is top soil and also would be impoverished after a year or two of cropping. (This sample was taken about one year after the garden had been abandoned following two years of cropping. Its poverty as compared to sample 3 cannot, however, necessarily be ascribed to gardening, since the sites are separated by more than 1,000 feet.)

Sample 42 is sufficiently rich to support a permanent agriculture if no water table or erosion problems exist. Sample 69 is also very rich. (Both of these samples are of surface litter, however, and are only 1.2 inches in thickness. It is doubtful whether this would be sufficient for permanent cultivation.)

Sample 74, which was taken from the B horizon in a garden abandoned two years earlier after two years of cropping, is very poor.

Sample 89 appears to be "the best general soil" for agriculture. Despite the fact that this sample came from the B horizon, it is quite high in nutrients and might support a "very good agriculture."

Dr. Popenoe concludes that "in general the soil samples appeared to be very low in nutrients and for these soils one might suggest that the crop rotation or swidden cycle helps to maintain the fertility of the soil for crop production."

APPENDIX 3

Floristic Composition of Primary Forest

LOW-ALTITUDE HIGH FOREST (*Wora Geni*), 2,200′–4,000′

Low-altitude high forest survives only in remnants. These are limited in size (the largest covering less than five acres) and are found only in special locations, such as on the crowns of knolls and on the saddles of ridges. Man's removal of the primary forest from adjacent areas has resulted in the increased penetration of sunlight into the lower strata of the remnants of this association, which has undoubtedly affected the floristic composition of these strata. It may be that the composition of the lower strata has also been affected by the absent-minded hacking of those who walk through with ax or bushknife in hand.

No forest census was undertaken in the low-altitude high forest. In one knoll-top location, 3.2 acres in area, the presence of "A" stratum trees was noted, however. These included six native taxa in at least four families: *nuŋ* (*Fagaceae, Quercus* sp.); *banč* (*Magnoliaceae, Elmerrillia papuana*); *kinde* (*Sapindaceae*); *aŋa* (*Sapotaceae, Planchonella* sp.); *tuem* (unident); and *dambi* (unident). Other common "A" stratum trees noted elsewhere in the low-altitude high forest included *Spondias dulcis, Pangium edule, Aleurites moluccana, Araucaria hunsteinii, Eugenia* spp., *Ficus* spp., and representatives of the *Lauraceae* family.

HIGH-ALTITUDE HIGH FOREST (*Kamuŋga Geni*), 4,500′–6,000′

The entire area above a line ranging from 5,000′ to 5,500′ is under unbroken high forest. Both the structure and composition of this association are easier to observe than in the lower altitude remnants. A forest census was undertaken at Tendopeŋ, altitude 5,000′, on a plot 200′x17′. The survey is summarized in Table 13. Because of the difficulties in securing adequate botanical samples from the large trees, most of the components of the A stratum remain unidentified.

Table 13. Survey of High Forest at Tendopeŋ
(Altitude 5,000′; Plot 3,325 sq. ft.; 200′ x 17′)

Native name	Spec. no.	*Circumf.	Lowest branch (est.)	Native category	Identification	Comments
			"A" Stratum			
dukumpina	256	6′2″	50′	apuŋ		
munduka	93	7′3″	60′	apuŋ	*Dilleniaceae, Dillenia* sp.	
yaŋgra		7′	70′	apuŋ		buttress roots
dumbi		3′	60′	apuŋ		
yimboka		5′6″	70′	apuŋ		buttress roots
yimuŋger		2′5″	60′	apuŋ		
dupai	381	5′6″	40′	apuŋ	*Lauraceae, Litsea* sp.	
miñiŋgambo	334	3′	50′	apuŋ	*Rubiaceae,* sp.	
yendek	286	6′	60′	apuŋ		buttress roots
Total "A" Stratum: 9 trees						
			"B" Stratum			
rama	143	3′	25′	apuŋ	*Urticaceae* sp.	
tondomane	76/84	1′8″	15′	apuŋ	*Theaceae, Ternstroemia* sp.	
ger	365	3′	25′	apuŋ	*Euphorbiaceae Codiaeum variegatum* (L)	
kina	223	2′6″	15′	apuŋ		
Total "B" Stratum: 4 trees						
			"C" Stratum (circumference over 6″)			
dimbi (2 spec.)		8″				
aimam	238	6″	15′	apuŋ	*Araliaceae, Boerlagiodendron* sp.	
wombo (2 spec.)		6″	15′	apuŋ		
ameŋgi	152A	6″	15′	apuŋ		
riŋganč (5 spec.)	238	12″	10′	apuŋ	*Moraceae, Ficus trachypison*	
tandomane	76/84	9″	12′	apuŋ	*Theaceae, Ternstoroemia* sp.	
dupai	381			apuŋ	*Lauraceae, Litsea* sp.	
koip	245	6″	20′	apuŋ	*Anacardiaceae, Semecarpus, prob. magnificans*	
rambai		6″		apuŋ		
da	404/312	6″	8′	apuŋ	*Lauraceae, Cryptocarya*	
Total "C" Stratum: 16 trees						

Table 13 (*continued*)

"D" and "E" Strata

Native name	Spec. no.	*Circumf.	Lowest branch (est.)	Native category	Identification	Comments
1. Other varieties of trees, circumference under 6″ (species found in "A," "B," and "C" strata not reported here even if present)						
tamp				apuŋ		
mambruŋ				apuŋ		
2. Shrubs, grasses, vines, etc.						
koriŋgi	109			pikai	*Urticaceae, Elatostema* sp.	ground shrub
anjomar	83			pikai	*Zingiberaceae, Alpinia* sp.	ground shrub
teraipind-pinda	134			bep	*Polypodiaceae, Dennstaedtia* sp.	ground shrub
kabaŋ bep	213			bep	*Cyatheaceae, Cyathea rubiginose*	supported on tree branches, but rooted in ground
doŋgai	137			bep	*Marattiaceae, Angiopteris* sp.	ground shrub
morameka	113/259			gawa	*Piperaceae, ?Piper* sp.	climbing vine
kriŋa	182			gawa	*Pandanaceae, Freycinetia* sp.	climbing vine
gañiŋgai				pikai	*Urticaceae, Elatostema* sp.	ground shrub
mopaka	263			gawa	?*Moraceae, Ficus* sp.	climbing on mundoka (Dillenia) climbing vine
močam	129			pikai	*Araceae, Aglaomena* sp.	ground shrub
tiwaka				gawa		climbing on dukumpina, climbing vine
terai	134			gawa	*Polypodiaceae, Dennstaedtia* sp.	fern vine, climbing on dukumpina

Table 13 (*continued*)

Native name	Spec. no.	*Circumf.	Lowest branch (est.)	Native category	Identification	Comments
kwiop	4/48			bep	*Polypodiaceae, Polypodium* sp.	growing on small rambai tree, a fern vine
yindim				bep		ground shrub
kwiopmai				gawa		climbing fern

* 3′ above ground, or above buttresses.

Other "A" stratum trees noted outside of the survey area included, in addition to many native taxa for which identifications could not be obtained, representatives of at least seven families. These are summarized in Table 14.

Table 14. Identified A Stratum Trees Noted Outside Survey Plot, Tendopeŋ

Family	Genus & species	Native name
Euphorbiaceae	*Macaranga* sp.	konjenipai
Fagaceae	*Quercus* sp.	nuŋ
Lauraceae	*Breilshniedia* sp.	kom
	Cryptocaria sp.	kawit
	Cryptocaria sp.	da
		boko
		gumbiaŋ
Magnoliaceae	*Elmerrillia papuana*	banč
Myrtaceae	*Eugenia* sp.	apeŋ
	Eugenia sp.	nonomba
Winteraceae	*Bubbia* sp.	ruiman

The floristic richness of the "B" and "C" stratum is also hardly illustrated by the census. Elsewhere between 5,000′ and 6,000′ representatives of the *Cunoniaceae, Guttiferae, Leguminosae, Melastomaceae, Myrtaceae, Palmae, Pandanaceae, Podocarpaceae, Rutaceae,* and *Saurauiaceae* families were noted, as well as additional native taxa among the *Euphorbiaceae, Lauraceae, Moraceae,* and *Urticaceae.*

Noted elsewhere in the "D" and "E," or ground, strata were members of the following families: *Balsamiferae, Compositae, Orchidaceae, Polygonaceae, Rubiaceae, Tiliaceae, Violaceae,* and *Umbellaceae,* as well as further representatives of families found on the survey plot.

Among large climbers noted between 5,000′ and 6,000′ were *Anonaceae, Pandanaceae, Rubiaceae,* and *Moraceae,* species. Strangling figs were prominent among the latter. The small climbers included members of the *Convulvulaceae, Ericaceae, Flagellariaceae, Gesneriaceae, Passifloraceae,* and *Rosaceae,* as well as the families found in the survey plot. Ferns, particularly *Asplenium, Lycopodium,* and *Nephrolepsis* are abundant among the epiphytes, as are several native taxa among the *Orchidaceae,* and epiphytic *Araceae* are present. Mosses cover the lower trunk of many "B" and "C" stratum trees and are present in patches on the trunks of the larger trees.

MOSS FOREST (*Kamuŋga Geni*), 6,000′–7,000′

Among the tallest trees in the moss forest are members of the *Lauraceae,* particularly species of *Litsea* and *Cryptocarya,* as well as *Eugenia,* with occasional individuals attaining an estimated height of 100′–125′. In addition to the *Lauraceae* and *Myrtaceae, Guttiferae, Logoniaceae, Melastomaceae,* and *Moraceae* are well represented among the trees. An *Astronia* (*Melastomaceae*) species is particularly abundant. A wild *Musa* sometimes achieves an estimated height of 60′ or 70′, and above 6,500′ two native taxa of *Pandanaceae,* probably separate species of the genus *Pandanus,* become plentiful. Both attain estimated heights of over 80′. *Elatostema* remains abundant on the ground, where representatives of the *Chloranthaceae* and *Marantaceae* families, as well as various ferns, particularly *Dawsonia* and *Polypodium,* are common. Vascular epiphytes and climbers are less common than in the high forest, but epiphytic mosses are much more luxuriant, completely covering the lower trunks of most trees to a depth of an inch or more.

APPENDIX 4

Estimating Yields per Unit Area

In the cases of both taro-yam and sugar-sweet potato gardens, figures are based upon the harvesting records of three gardens. This was necessary because the period of fieldwork, fourteen months, was considerably shorter than the life of any garden, figured from planting to abandonment. Moreover, harvesting records could not be started immediately upon arrival among the Tsembaga: it was necessary first to win the confidence and cooperation of the people. Harvesting records thus began on February 14 and ended on December 14, 1963. There was, therefore, a two-month period of the year for which no records were compiled.

For the first twenty-three weeks of both the taro-yam garden and the sugar-sweet potato garden harvesting figures from the garden of Moramp and Mer at Torpai (3,900′–4,000′) were used. This was a mixed garden, resembling taro-yam gardens more closely than sugar-sweet potato gardens. It would have been preferable to use, for the sugar-sweet potato gardens, figures from a true sugar-sweet potato garden but, as has already been mentioned, no true sugar-sweet potato gardens were being made in 1963 because the reduction of the pig herd was imminent. However, the expedient settled upon—that of using the Torpai garden figures for both types of gardens—is not misleading. The two are very similar until the tubers ripen. Some adjustments were made, however. While all of the Torpai harvesting figures were used directly for the taro-yam garden, some of them were halved for the sugar-sweet potato garden computation. These included the figures for *kwiai* (*Setaria palmaefolia*), *ira* (pumpkin), *yibona* (gourd, ? *Lagenaria*), *čeŋmba* (*Rungia klossi*), and miscellaneous greens. It is perhaps the case that a further adjustment should have been made for *pika* (cucumber). In any event, these are low-calorie foods and, with the exception of cucumber, the yields were small. For the period from the twenty-fourth week after planting until the sixty-sixth week, figures are based, for the taro-yam garden, on the garden of Moramp and Mer at Kakopai (3,800′), and, for the sugar-sweet potato garden, on that of Walise and Pambo at Timbikai (4,400′–4,500′). Both of these gardens were planted late in the 1962 planting season, and harvesting records exist from the first extraction of the root crops. The Torpai records also include the first extraction of root crops, so it was possible to fit both the Timbikai and Kakopai records to the Torpai record. It would seem, to put it a little differently, that both Timbikai and Kakopai were planted in late August or early September of 1962, while Torpai was planted in late June of 1963. The nine-week gap thus disappears.

The period from the sixty-seventh week after planting until abandonment is represented, in the case of the taro-yam garden, by the garden of Walise and Pambo at Tipema (4,000′–4,100′), and for the sugar-sweet potato garden by that of Ačimp and Avoi at Unai (4,100′–4,200′). In gardens of this age it is not possible, however, to discover signs indicating the age to within a few weeks, and it may be that nine weeks or more of harvesting figures are lost. The gardens, that is, may have been not sixty-seven but seventy-six or even more weeks of age when records commenced. It is quite clear that they were not under sixty-seven weeks' old when record keeping began because of the great disparity in root crop production in the last month for which records were kept on Kakopai and Timbikai, and the first month for which records were kept on Tipema and Unai. (The per-acre sweet potato yield, for instance, for the last month of Timbikai production was 638 pounds. The first month of Unai records showed a yield of only 158 pounds per acre.) This is not an infallible indication, for sweet potato harvesting drops off quite rapidly in any garden when a younger garden begins to produce in quantity. Unfortunately, the data are not sufficient, because of the nine-week hiatus, to know exactly how rapidly it does decline.

Despite the possibility of the record being short of some weeks of harvesting figures, no adjustments were made. The Unai and Tipema figures were accepted, that is, as if they did in fact represent harvesting between the sixty-seventh and one-hundred-tenth weeks after planting. This has probably resulted in an underestimate of the production for the period. The magnitude of such error, however, in comparison with the total yield of the garden during its entire life cannot be too great.

An additional underestimation results from the fact that I left the field before the gardens were completely abandoned. Some sugar cane and bananas were surely harvested after December 14. Again, the amounts were without doubt insignificant compared to the total harvested during the earlier life of the gardens.

These possible underestimations in regard to the harvesting from gardens over sixty-seven weeks of age are, in a sense, corrective. Both Tipema and Unai are on the government walking track and convenient to later gardens of both Ačimp and Avoi on the one hand, and Walise and Pambo on the other. They were, therefore, during the period after the sixty-seventh week, harvested more intensively and for a longer period than would have been the case had they been in less convenient locations.

Additional errors may have been introduced by some of the adjustment factors employed. These include factors for nonweighed harvesting, percentage of edible portion, and calories per pound of edible portion. All weighing took place at the women's houses. Some food, however, was consumed without being brought home. Bananas, for instance, are a frequent noon-time snack, and often a tuber is thrown on a fire in the garden at mid-day. On some occasions, moreover, an earth-oven is made in the garden and a full meal is taken there.

Most importantly, the "refreshers," sugar and cucumber, are eaten in the garden during the working period. It was my impression that twice as much sugar was consumed away from the women's houses, where weighing took place, as at the houses. This impression is based upon observations made without a scale and may be high.

Finally, some foods, notably bananas, sugar, *pitpit,* and greens, were sometimes brought directly to men's houses for consumption, without having been first weighed.

It will be noted that the factors for nonweighed harvesting differ between the earlier stages and the final stage of the gardens. This is because much of the consumption represented by nonweighed harvesting occurs in the course of work, and little work takes place in old gardens.

The figures for edible portions were worked out by tests in the field. The samples were small, and there are considerable problems involved in working out edible portions. Even within single varieties of single species there is variation between specimens of different sizes, ages, etc. The edible portion factors determined by tests, however, accorded fairly well with figures found in the literature.

Similar problems may be raised concerning the caloric values assigned to the various crops. These, too, vary within single named varieties of different ages, sizes, or origin. Caloric values, like other food values, were taken from a literature in which there is often wide disagreement. Where it was possible, the values adopted were those used in other studies, notably that of Hipsley and Clement (1947) in nearby or similar environments. One of the chief virtues is comparability. Sources for the values are listed in Appendix 9.

All gardens have some idiosyncratic features. The gardens from which yield figures were derived are no exception. Tipema, which represents the sixty-seven to one-hundred-tenth weeks of the taro-yam garden, happened to be totally without *Xanthosoma,* which is generally planted in wet areas at the bottom of slopes. At Tipema, however, the government walking track forms the lower margin of the garden and acts as a kind of drainage ditch, rendering an area that would ordinarily be planted in *Xanthosoma* unsuitable for this crop. It should also be mentioned concerning Tipema that although the amount of *kwiai* present was extraordinary, it may have been further overestimated. Such an overestimation would make little difference in the total caloric figure, however.

It was already mentioned, in respect to the sugar-sweet potato gardens, that cucumber may be overestimated. A further inaccuracy, or inconsistency at least, lies in the banana figures. While it is the case that more bananas are harvested after the sixty-seventh week than before it, the ratio in most cases would not approach the approximately 5.5 to 1 indicated here. With increasing altitude, fewer bananas are planted and maturation is slower. Timbikai, which is about 300′ higher than Unai, exemplified this general trend.

APPENDIX 5

Energy Expenditure in Gardening

METHOD

For the purpose of estimating the amount of energy expended in producing an acre of garden, persons of known body weight were observed in the performance of each of the gardening procedures. The amount of time necessary to accomplish the various tasks was recorded. This was converted directly into time per unit area, in the case of garden making. In the case of harvesting it was converted first into time per weight of harvested material. Since yields per unit area were known, it was then possible to convert the time per weight figures into time per unit area figures.

Since the body weights of all persons observed were known, a caloric expenditure per unit of time could be assigned to their efforts in various tasks by reference to *Rates of energy expenditure under varying conditions of activity and body size (adults)*, a table compiled by Hipsley and Kirk (1965:43) upon the basis of observations and gas exchange measurements made among the Chimbu, a highland New Guinea people. This table summarizes caloric expenditure per minute for persons of various body weights and ages performing twenty-six different tasks.

It could not be assumed, of course, that corresponding procedures of Tsembaga and Chimbu are equally strenuous. Before values from Hipsley's and Kirk's table could be adopted, therefore, evaluations had to be made of the level of effort required in the performance of the various tasks. Such an evaluation was arrived at by counting the number of hand movements per minute and by considering other factors such as the bulk and weight of the materials manipulated. This usually permitted the acceptance of values for particular procedures determined by Hipsley, but in some cases direct adoption was not warranted. Fence making, for example, apparently is not particularly heavy work among the Chimbu, who use much lighter timber than do the Tsembaga. Observation of the procedure among the Tsembaga led me to select Hipsley's and Kirk's values for the heaviest work rather than the value they present for fencing.

PROBLEMS WITH THE METHOD

It may be questioned whether the application to the Tsembaga of figures compiled upon observations of the Chimbu is justified. It may be noted that the Chimbu live and work in a higher, cooler environment than do the Tsembaga. No adjustment has been made for temperature differences, and it may be that the caloric expenditure values assigned to the Tsembaga for various tasks are therefore slightly high.

The Chimbu, moreover, are larger people than the Tsembaga. While Hipsley's and Kirk's figures do include values for body weights within the Tsembaga range, it is possible that some further adjustment would be desirable.

The nutritive statuses, furthermore, of the Chimbu and Tsembaga differ. Venkatachalam (1962:10ff) indicated that the Chimbu suffer more severely from certain dietary deficiencies than do the Tsembaga; these deficiencies may have affected the values that Hipsley and Kirk derived.

Despite these difficulties, the use of values derived from observations made upon another New Guinea population seemed preferable to the use of less detailed tables based upon observations of European populations. It may be suggested, however, that anthropologists who are concerned to estimate energy expenditure should, if at all possible, arrange to have tests made upon the populations with which they are working.

INTER- AND INTRAINDIVIDUAL CONSISTENCY

An additional problem is raised by the possibility that the ranges of performance by various Tsembaga in the several tasks involved in gardening were greater than were reflected by the samples upon which evaluations were based, since samples were small. In the case of fence making, evaluations are based upon observations of only two workers, for example. For other activities the samples were larger, although always of limited size. There are, however, indications that differences in the performances of various actors or single actors at different times were not great. This is most clearly exemplified in the clearing of underbrush, a highly standardized task, since the composition and density of herbaceous undergrowth is similar on most plots in which gardening takes place. The operator removes this material by both uprooting it and by cutting it as close as possible to the surface of the ground with a bushknife. Only on sites on which *kunai* (*Imperata cylindrica*) is prevalent is the procedure different. Because its subsurface structure is rhizomatous, and because the edges of its leaves are razor sharp, no attempt is made to uproot this grass. Where it is plentiful the gardener simply cuts as close to the surface as possible. Such sites are rare, however; on only 6 out of 381 sites examined was *kunai* common. Seven individuals were observed clearing underbrush on three separate days. The results are summarized in Table 15.

Table 15. Clearing Underbrush—Time and Motion Study

Worker's name	Sex	Est. age	Weight in lbs.	Time	No. strokes	Time	No. strokes	Time	No. strokes	Comments
Akis	M	20	88	10:37–10:43	296	11:14–11:20	250	12:14–12:20	248	Only one 3-minute break during period. Next longest break: 15 seconds.
Ačimp	F	50	85	10:55–11:01	244	11:30–11:36	209			No breaks longer than 20 seconds during working period.
Avoi	M	55	94	11:02–11:08	177	11:50–11:56	190			Slower than other workers because of short breaks, and slower strokes.
Meñ	M	28	120	6 min.	233					Longer strokes than any of the others.
Wale	F	35	76	9:53–9:59	246	10:58–11:04	260			No breaks longer than 20 seconds during working period.
Nimini	M	18	96	6 min.	246					
Mer	M	40–45	94	6 min.	316					Stated that he was in a hurry.

Several remarks are in order concerning the results of this limited sampling. First, if the performances of Avoi, who is elderly and frail, and of Mer, who was by his own statement hurrying, are eliminated, the tempo of performance by various actors is quite uniform. Second, the tempo of single actors through time is also quite consistent. It may be mentioned that periods of rest during the course of the work are uncommon, except for a protracted break around mid-day.

In the case of only one of the workers, Wale, was the area cleared actually measured by chain and compass. Over a period of two hours Wale cleared at the rate of 210 square feet per hour. Wale, it may be pointed out, is a very small woman, and the larger, stronger men are able to clear larger areas in the same amount of time. I estimated by eye that Nimini cleared about 250 square feet an hour, and that Meñ, an exceedingly muscular man, was working at about the rate of 300 square feet an hour.

A comparison of the energy performances of Wale, Nimini, and Meñ may be made by referring to the energy expenditure values assigned to various activities by Hipsley in his study of the Chimbu. For an individual of Wale's size, medium-heavy garden work, such as the clearing of underbrush, requires the expenditure of 2.35 calories per minute or 141 calories per hour over basal metabolism. A person of Nimini's size, on the other hand, expends energy at the rate of 2.62 calories per minute, or 157.20 calories per hour. At these rates Wale is expending .67 calories to clear one square foot of underbrush, and if the estimate of 250 square feet per hour in Nimini's case is accurate, he is expending .63 calories per square foot. Meñ expends energy at the rate of 3.28 calories per minute over basal metabolism in this task. If the estimate that 300 square feet are cleared hourly by Meñ is accurate, he is expending .66 calories per square foot.

Harvesting of sweet potatoes shows similar consistency in energy expended per pound of tubers. In a garden about one year of age, Koi, 4′ 5½″ tall and weighing 87 pounds, was timed over a period of 85 minutes. Her work involved about 75 hand movements per minute during the entire harvesting period, and yielded 23 pounds, 2 ounces of tubers. The value assigned her efforts were 1.60 calories per minute, or 5.9 calories per pound of tubers harvested. Ačimp and Walise were also observed in sweet potato harvesting. Their efforts showed slightly poorer results: 6.3 and 6.9 calories expended per pound of sweet potatoes harvested. The tempos at which they worked were similar to that at which Koi worked, but the gardens were somewhat older, fourteen or fifteen months, than the year-old garden in which Koi was observed. Yields were perhaps beginning to fall off slightly. Informants say that in younger gardens return for work is greater, but no figures were obtained, unfortunately, for sweet potato harvesting in gardens under one year in age.

The data are insufficient, but it may be that the evenness of the tempo both through time and between individuals represents an approach to some sort of energy optimum in which accomplishment (e.g., square feet of land cleared) is maximized in relation to energy expenditure. A much larger set of observations on persons representing a full range of body sizes would be a worthwhile undertaking. It is possible that such a study would not only yield insights into optimal tempos for particular tasks, but also optimal body sizes for particular tasks or complexes of tasks. It may well be, for example, that the extra energy expended by a 150- or 160-pound man in clearing underbrush would not be matched by additional accomplishment. The caloric expenditure, that is, might rise well above .66 per square foot. This might illuminate a possible selective factor that has generally been ignored in discussions of evolution and might have particular relevance to pygmy populations.

COMPARISON OF CROPS

It may be mentioned that the estimation of the energy expenditure in various procedures involving different crops provides an additional dimension for comparing their desirability to the gardener. For example, the cost of harvesting sweet potatoes, which we take to be 5.9 calories per pound, may be compared to the cost of harvesting *Xanthosoma,* only 1.1 calories per pound. Using a value of 681 calories per pound for sweet potatoes (see Appendix 9) and 658 calories per pound for *Xanthosoma,* the ratio of gross energy yield to energy input in sweet potato harvesting is 116:1, while the ratio of gross energy yield to return in *Xanthosoma* harvesting is 598:1. If corrections are made for the yield of tubers fit for human consumption, the disparity between sweet potatoes and *Xanthosoma* is even greater. Thirty to fifty percent of the sweet potatoes are of small size (under four ounces), which, while edible, are considered unpalatable, being mostly skin. Small *Xanthosoma* tubers, having a thinner skin and a rounder shape, are considered to be quite acceptable.

There are at least thirty-eight named varieties of *Dioscorea* in five different species grown in Tsembaga gardens, and there is a considerable range in the depth at which the tubers mature. The extraction of some types requires that holes three to four feet deep be dug, and even the shallow varieties demand excavations of twelve to eighteen inches for their harvesting. While in some instances fifteen to twenty pounds of tubers may be obtained from a single plant this is rare, and it is clear that in terms of energy expenditure the harvesting of yams is less rewarding than is the harvesting of any other root crop. The figure of ten calories per pound, an estimate, is probably too low. The advantage of the *Dioscorea,* however, aside from their storageability, is just that they do mature at greater depths than do other root crops. Forming a deep-lying stratum of edible material that does not impede the development of other more shallow strata, they provide the gardener with the opportunity to increase his per-acre yield. This advantage of *Dioscorea* may be sufficient to offset the costliness of harvesting them.

It may be mentioned that it is not only in harvesting that various crops may have differing energy characteristics. In mixed swiddens of either the taro-yam or sugar-sweet potato types weeds are much more of a problem than they are in the single species stands of *Xanthosoma,* which are planted at the lower, wetter margins of gardens of both kinds. The broad leaves of these plants generally shade out any other species that may attempt to establish themselves in their immediate vicinity. When the size of the energy expenditure in weeding is examined, this difference is seen to be of considerable importance.

ENERGY EXPENDITURE IN VARIOUS TASKS

The clearing of underbrush requires considerably more effort than the clearing of second-growth trees. Before steel was introduced, moreover, the disparity between energy expenditure in these two operations must have been greater, for while stone axes are quite efficient for felling second-growth trees they are poor implements for clearing underbrush. Before the introduction of the bushknife, underbrush was cleared by laying a pole three or four inches in diameter on the ground, pulling the weeds over this pole and chopping them with an ax. The indication is that the bushknife, at least as far as reduction of energy expenditure in gardening is concerned, was a more important introduction than the ax. (The ax is additionally more useful in chopping firewood.)

The amount of energy expended in fencing, about 46 calories per linear foot, goes far to explain why the cultivations of a number of gardeners are clustered, and the cost of transporting foodstuffs from the gardens to the houses provides a strong indication as to why the scattering of gardens as a result of the increasing size of the pig herd is accompanied by a scattering of residences. As noted in Table 5, a reduction of garden-to-house food transportation by 80 percent would improve the energy ratio of sugar-sweet potato gardens from 15.9 to 18.4:1, and that of the taro-yam gardens from 16.5 to 20.1:1. It should be mentioned that it is not only food that is carried from the gardens to the houses but also firewood, for swiddens are the most important sources of this important commodity. (When trees are cut to make a garden some of the logs are split and stacked to dry for fuel. After the abandonment of gardens, pollarded trees are frequently felled for fuel.)

COMPARISON WITH OTHER AREAS

Harris, in the unpublished paper referred to earlier, has assembled estimates of the energy ratios characteristic of several nonmechanized agricultural systems. The Tsembaga ratios, 16.5 to 20.1:1 and 15.9 to 18.4:1, although high, fall within an expected swiddening range. Harris' estimates for swiddening in other areas include Dyak (Borneo) rice swiddens, 10:1, and Tepotzlan (Yucatan) maize swiddens, 13:1 on poor land and 29:1 on good land.

Other agricultural systems may be much more efficient: Harris estimates, on the basis of Fei's (1945) data, that energy ratios in wet rice paddies reach 53.5:1 in Yunan. Harris points out that wet rice agriculture is not always so productive, however. Gerieri (Gambia) swamp rice cultivation shows an energy ratio of 11:1 (Haswell, 1953), which is about the same as the energy ratio (10.7:1) derived for their "savannah hoe cultivation" of other grains. Dyak wet rice cultivation, according to Harris, shows an energy ratio of 14:1.

Furthermore, a direct comparison of Tsembaga and Yunan figures may be misleading. Well over 90 percent of the Tsembaga diet is taken from the swiddens, and the energy expenditure figures presented in Table 5 would include some casual hunting. On the other hand, rice accounts for only 70 to 80 percent of the Yunan diet, and we have no estimate of the efficiency of the processes involved in other aspects of the food quest.

Despite the Yunan estimation, it is possible to say of Tsembaga gardening that although it seems to be characterized by comparatively low yields per unit of area, it shows at the same time comparatively high yields per unit of energy input. This characteristic of Tsembaga swiddening accords well with another of its aspects, i.e., that of disturbing as little as possible the secondary forest community in which the gardening takes place.

High return on energy input may be a characteristic of swiddening in general. Concerning the shifting cultivation of maize by the Amahuaca in eastern Peru, Carneiro says, "All in all, then, Amahuaca shifting cultivation is, in spite of its rudimentary form, thoroughly capable of producing the food abundantly, reliably, and with relatively little expenditure of labor" (Carneiro, 1964:18). Concerning Hanunoo dry rice cultivation, Conklin (1957:152) suggests an adjusted value of 2.5 kilograms of unhusked rice (caloric value 3,600 calories/kg.) as the return per man hour of labor. He notes, "This rough estimate compares favorably with labor cost figures for rice production under the best conditions elsewhere in the tropics." Both Carneiro's and Conklin's statements are, admittedly, based upon only partial quantification of the relevant variables, but both are sophisticated and experienced fieldworkers.

APPENDIX 6

Secondary Growth

Samples of second-growth composition (*riŋgopwai*) were taken by chain transit in six producing gardens between 3,400 and 4,800 feet in altitude. A total of twenty-one arboreal species was noted, with four to ten species in individual gardens. The most common species were *gra* (*Dodonaea viscosa*), specimens of which were in total the most numerous, but which were represented in only four out of six of the gardens, and *pokai* (*Alphitonia iacana*), somewhat less numerous than *gra* but appearing in all gardens. Also common was *gapni* (*Homolanthus* sp.), which appeared in five out of six gardens and was almost as numerous as *pokai*. These three species accounted for 43 percent of the sample, which included 117 specimens. Table 16 summarizes the composition of the sample. At least eight of these trees, *kobenum, kamakai, ganč, gonwant, gum, noŋ, riŋganč,* and *yiŋgra,* also occur in the primary forest.

There are some differences between gardens in the composition of the arboreal component of their early fallow associations, but it is not possible for me to state to what extent these differences are a function of altitude or of other factors, such as local edaphic conditions, proximity to virgin forest, or length of the previous fallow. Figures in Table 16 should not be taken to indicate differences in the prevalence of trees in the various gardens; there are differences, but they are not reflected in the sample.

The speed with which individual trees grow, and the prevalence of trees on the plot varies, roughly, inversely with altitude. More second-growth trees appear on lower altitude gardens and grow faster than in the higher altitude gardens, although this correlation is sometimes masked by local conditions. In one garden at 3,600 feet not included in the sample, the average height of second-growth trees was estimated, less than eighteen months after planting, to be twelve to fourteen feet, and they

Table 16. Trees Appearing in Tsembaga Gardens prior to Their Abandonment

Trees			Gardens and Altitude						
Family	Genus & species	Native name	Porakump 3,400′	Tipema 4,000′	Tapipe 4,100′	Unai 4,400′	Timbikai 4,500′	Bokandipe 4,800′	Total
Cunoniaceae	*Caldcluvia* sp.	bokanč					2		2
Cyatheaceae	*Cyathea angiensis*	yimunt		3				6	9
Euphorbiaceae	*Homolanthus* sp.	gapni	1	2	7	1	1		12
Fagaceae	*Quercus* sp.	noŋ						1	1
Guttiferae	*Garcinia* sp.	gum						1	1
Melastomaceae	*Melastoma malabathricum*	wopkai		2			1		3
Moraceae	*Ficus puncens*	kobenum					3	2	5
	Ficus trachypison	riŋganč				1			1
Rhamnaceae	*Alphitonia iacana*	pokai	3	1	2	3	4	2	15
Rubiaceae	sp.	ganč		2	1				3
Sapindaceae	*Dodonaea viscosa*	gra	6		1	3	12		22
Saurauiaceae	*Saurauia* sp.	goŋgo		3			1		4
Solanaceae	*Solanum* sp.	gon		3				7	10
Sterculiaceae	*Colona scabra*	kamkai		2	1			2	5
Ulmaceae	*Gironniera* sp.	penda				1	1	2	4
	sp.	dima		2					2
Urticaceae	*Procris* sp.	pentapent			2				2
Verbenaceae	*Geunsia farinosa*	gonwanč						1	1
Unidentified		beknan		2					2
"		mopo					1	2	3
"		yaŋgra	1		5		3		9
Total specimens in sample			11	22	19	9	29	26	116
Total named types represented			4	10	7	5	10	10	21

were close enough together for their crowns to have almost formed a continuous canopy. This garden, except for some bananas that remained to be harvested, had been abandoned between fourteen and sixteen months after planting. It seems probable that it was the density of the secondary growth that forced this early abandonment. Secondary growth in the higher altitude gardens seems to achieve somewhat less height and considerably less density in similar lengths of time. It is perhaps for this reason that they usually remain in production longer; sugar and *pitpit*, as well as bananas, are taken in some cases more than twenty-four months after planting in gardens between 4,500 and 5,000 feet in altitude.

While the development of the arboreal component of the secondary growth may be a direct function of altitudinal differences with attendant variations in temperature and edaphic conditions, it might also be a result of the difference in planting patterns at different altitudes. In the higher altitude gardens there is heavier planting in sweet potato. The presence of sweet potato is said to inhibit the growth of weeds, and it may inhibit the growth of young trees as well. Moreover, the techniques for harvesting sweet potato may result in the accidental removal of young trees.

The floristic composition of the herbaceous component of the second growth previous to the abandonment of the gardens shows about the same richness but somewhat less variation among gardens than the arboreal component. The composition of the sample collected by the same chain transits is summarized in Table 17.

The sampling technique is less likely to reveal uncommon herbaceous types than arboreal types, and the overall floristic composition of the herbaceous component of early secondary growth must be richer than the

Table 17. Herbs Appearing in Tsembaga Gardens prior to Their Abandonment

Trees			Gardens and Altitude						
Family	Genus & species	Native name	Porakump 3,400′	Tipema 4,000′	Tapipe 4,100′	Unai 4,400′	Timbikai 4,500′	Bokandipe 4,800′	Total
Balsaminaceae	*Impatiens platypelia*	korambe		1				1	2
Compositae	*Blumea balsamifera*	kwirañ	1	25	10	19	16	13	84
	Microglossa pyrofilia	riŋgop gaua		1					1
Cyperaceae	*Scleria* sp.	riai	1		4		1	1	7
Gramineae	*Coix lachrimajobi*	koŋgun	1	1	12		18	2	34
	Imperata cylindrica	korndo	1	2	1	1		4	9
	Isachne myosotis	piŋgo	1					10	11
	Ischaemum digitatum	bombak			1	1	3		5
	Paspalum conjugatum	tamo		4	7		4		15
	Phragmites karka	yamboč					2		2
	Polytoca aerophyla	wandama					8		8
	Setaria palmaefolia	korami	3	8	12	4		1	28
Orchidaceae	sp.	tiokum	1		1				2
Rosaceae	*Rubus moluccanus*	kurkur			1		2		3
Ferns	*Cyclosorus ?truncatus*	aruk		3		3			6
	Diplazium sp.	raŋgilopa				1			1
	Nephrolepsis schlechteri	nomapunt-mai		1					1
	Pteris sp.	bor		10		12			22
	Pteris sp.	kembor	6	2	1	2			11
Total specimens in sample			15	58	50	43	54	32	252
Total named types represented			8	11	10	8	8	7	19

sample reveals. The predominant species, however, are certainly represented. Three types that did not show up should be mentioned. These are *ambek* (*Solanum nigrum*), *mañump* (*Cyathea* sp.), and *kaŋgup* (*Cyathea* sp.). *Ambek* appears very early in newly burned gardens and, being edible, is harvested. *Mañump,* an edible fern appearing in gardens below 4,000 feet, is also harvested. *Kaŋgup,* also an edible tree fern, appears in gardens as well as *riŋgopwai* above 400 feet, is also allowed to mature, and is eventually harvested. *Kaŋgup* and *mañump* are protected, particularly when they appear in pandanus groves.

Edible mushrooms of several named varieties may also be considered to be part of the fallow, appearing on rotting logs and stumps of the cut *duk mi.*

Kunai (*Imperata cylindrica*), a notorious pest, is relatively rare in the sample. Prevalence of this grass on a site indicates deflection from a succession leading to an arboreal climax to a succession leading to a stable grassy disclimax, which it dominates. Its rarity in Tsembaga gardens indicates that such a disclimax is not being induced on the sites sampled. Less precise observations made on other gardens indicates that what is true for the sample is true for Tsembaga gardens generally.

The floristic richness of the *riŋgopwai* association increases rapidly with age. A survey of an area of several acres at 4,800 feet, part of which had been planted three and a half, and part four and a half, years previously, showed 118 native named types, most of which represented species. All the arboreal species present were probably reported, but it is possible that a good many less obvious herbaceous species were not.

The arboreal component of this association consisted of at least thirty-six named native types in thirty-two genera and twenty-two families. At least eighteen of these types are also found in the virgin forest; it thus seems that at an early age secondary growth already begins to bear a floristic resemblance to the virgin forest characteristic of its altitude. A contrasting survey at lower altitudes of sites that had lain fallow for a similar length of time was not conducted, however. The composition of the arboreal component of this association is presented in Table 18. Many of the trees on this site were twenty feet high, and diameters estimated at three to four inches were the rule. They were scattered over the area, their crowns not touching. This site was close to the settlement, however, and it is likely that the formation of a canopy was being suppressed by pigs, who in their rooting continually destroyed the seedlings appearing in the spaces between the well-established specimens.

A census of trees on an 11,010-square-foot site at an altitude of 4,200 feet, which had last been cultivated twenty to twenty-five years earlier

Table 18. Arboreal Component, Secondary Growth (Pra and Gerki, 4,800′–5,000′, Two to Three Years after Abandonment)

Family	Genus & species	Native name
Dilleniaceae	*Dillenia* sp.	munduka
Ebenaceae	*Diospyros* sp.	wonom
Ericaceae	*Rhododendron macregorii*	mer
Euphorbiaceae	*Breynia* sp.	nonmanč
	Glochidion sp.	mbanmban
	Homolanthus sp.	gapni
	Macaranga sp.	apapa
	Mallotus sp.	goŋgenaga
	Phyllanthus sp.	yuarundo
Fagaceae	*Quercus* sp.	noŋ
Guttiferae	*Garcinia* sp.	gum
Leguminosae	*Desmodium sequax*	koraindindiye
Moraceae	*Ficus* sp.	baŋgambai
	Ficus sp.	danje
Myrtaceae	*Cleistanthus* sp.	timbi
	?*Decaspermum necrophyllus* *Octamyrtus durmanni*	jijimbint
	Eugenia sp.	aŋkunung
	Eugenia sp.	tandapa
Palmae	*Licuala* sp.	morapmai
Podocarpaceae	*Podocarpus* sp.	minjaun
Rubiaceae	*Mussaenda pondosa*	goimbambo
	Psychotria sp.	yuaroro
Sapindaceae	*Dodonaea viscosa*	gra
	Mischocondon sp.	birpi
Saurauiaceae	*Saurauia* sp.	rokunt
Solanaceae	*Lycianthes* sp.	kapaŋ
	Solanum sp.	gon
Theaceae	*Ternstroemia* sp.	tondamane
Thymelaeaceae	*Phaleria nisidae*	pukna
Ulmaceae	*Gironniera* sp.	penta
Urticaceae	?*Leucosyke* sp.	naŋgrek
	Maoutia sp.	noŋgamba
	Procris sp.	pent pent
Verbenaceae	*Geunsia farinosa*	gonwant
Unidentified		kumpnai

Total tree species represented: 35

and which was about to be cultivated again, is summarized in Table 19. This association seemed less varied than the three- to four-year-old association discussed above. Only twenty-six native named types were noted. Resemblance to virgin forest, however, seemed to have increased; at least eighteen of the twenty-six are also found in the *geni*.

The large number of the tree fern *yimunt* (*Cyathea angiensis*) is unusual. Single species stands of *yimunt*, or associations approaching single species stands, are regarded by the Tsembaga as indicative of poor soil. The gardeners, however, informed me that the soil was good on this site, and soil sampling showed no special deficiency. There are indications that at an earlier stage in the fallow development on this site, *Cyathea* shared dominance with the quick-growing types *gra* (*Dodoneae viscosa*) and

Table 19. Census of Trees, Twenty- to Twenty-Five-Year-Old Second Growth (Tipema, Altitude 4,200′; Plot 11,010 Square Feet)

Family	Genus & species	Native name	No. spec. above 6″ circumf.	Largest spec. circumf. at 3′	Largest spec. height (est. feet)
	Specimens 6″ circumferences or larger				
Casuarinaceae	*Casuarina papuana*	ndumi	4	3′	60–70
Cunoniaceae	*Caldcluvia*	bokanč	2	1′6″	30
Cyatheaceae	*Cyathea angiensis*	yimunt	54	1′8″	25
Ebenaceae	*Diospyros* sp.	wonum	4	1′5″	30
Euphorbiaceae	*Homolanthus* sp.	gapni	1	1′4″	25
	Mallotus sp.	gimbint	2	9″	20
Moraceae	*Ficus puncens*	kobenum	4	7″	25
	Ficus trachypison	riŋganč	6	1′	20
Myrtaceae	*Eugenia* sp.	tandapa	1	6″	12
Ochnaceae	*Schuurmansia meningsii*	arare	1	1′3″	20
Rhamnaceae	*Alphitonia iacana*	pokai	3	2′6″	50
Rubiaceae	*Psychotria* sp.	burai	1	4″	20
Sapindaceae	*Dodonaea viscosa*	gra	4	1′	35
Saurauiaceae	*Saurauia* sp.	gaŋgo	15	2′6″	35
Solanaceae	*Solanum* sp.	gon	2	2′	35
Urticaceae	*Missiessya* sp.	yamo	5	1′3″	25
		rama	2	10″	20
Unidentified		mopo	4	2′6″	35
"		kariŋanč	1	3′1″	50
"		marmar	1	7″	15
Total specimens 6″ circumference or larger: 117					
	Other named types present, under 6″ circumference				
Melastomaceae	*Melastoma malabathricum*	wopkai			
Moraceae	*Ficus calopilinia*	muruŋga			
	Ficus wassa	kundua			
Unidentified		dukumpina			
"		punt			
"		raŋgan			
Total named tree types: 26					

pokai (*Alphitonia iacana*), which were, at the time of the census, quite obviously being suppressed. Two of the four *gra* present were moribund.

The structure of this twenty- to twenty-five-year-old secondary forest differed markedly from the structure of the three- to four-year-old association. The sizes of individual specimens are indicated in Table 19. The crowns of the larger trees formed an unbroken canopy over the entire area, so that the ground layer of shrubs, grasses, and creeping vines was, except for an occasional sunbeam, in shade.

A further idea of the association is perhaps conveyed by the density of trees over six inches in circumference: there were ninety-four square feet per specimen. This seems typical. In intervening spaces, in addition to the sapling types listed in Table 19, rather large *yikon* (*Piper* sp.) bushes, or shrubs, were present on the site. Density of undergrowth was not, however, sufficient to impede rather extended visibility five to six feet above the ground.

APPENDIX 7

Commonly Propagated Plants

Table 20. Commonly Propagated Plants, Tsembaga

Family	Genus & species	Native name	Life form	Use	Where planted
Acanthaceae	*Graptophyllum* sp.	yenɟim	bush	landscaping	at houses and raku
Anacardiaceae	*Mangifera indica*	wowi	tree	food (fruit)	wora
Araucariaceae	*Araucaria hunsteinii*	yuk	tree	house protection	At houses and raku
Cruciferae	*Roripa*[a] *Nasturtium aquaticum*	gonbi	herb	food (leaves)	shallow stream beds
Euphorbiaceae	*Aleurites moluccana*	kaba	tree	food (nuts)	close to houses in wora
	Codiaeum variegatum	ger	bush	boundary mark	on boundaries
Gnetaceae	*Gnetum gnemon*	ambiam	tree	food (leaves)	wora
Gramineae	*Bambusa forbesii*	kinjen	reed	roofing (leaves)	kamuŋga-wora amaŋ & wora
	Bambusa sp.	muŋ[b]	bamboo	building	near houses
Leguminosae	*Erythrina* sp.	yaur	tree	food (leaves)	near houses
Liliaceae	*Cordyline fruticosa*	rumbim[c]	small tree	boundary marker and ritual	boundaries, raku, houses
Moraceae	*Ficus wassa*	beka (kundua)	tree	food (leaves, fruit), fiber, pith	wora
	Artocarpus sp.	mokoi	tree	Food (fruit)	below 3,800′
Myrtaceae	*Cleistanthus* sp.	timbi	tree	ritual	near wora raku
Palmae	sp.	bina	tree	bows, arrows, spears	wora
Thymelaeaceae	*Phaleria* sp.	pukna	tree	fiber, pith	near houses
Urticaceae	*Oreocnida* sp.	rumen	bush	forms hedge	house, raku, boundaries
Zingiberaceae	*Amomum* cf. *polycarpum*	gunuma	reed	ritual, food (fruit)	boundaries, houses, raku
Unidentified		tup	bush	dye, red	near houses
"		kalom			
"		tup	bush	dye, black	near houses
"		ruŋgi			

[a] European watercress introduced ca. 1957.
[b] Many named types, probably species.
[c] Many named types, horticultural varieties.

APPENDIX 8

Nondomesticated Resources

The following lists of nondomesticated materials used by the Tsembaga cannot be regarded as exhaustive. Each item is listed in the association in which it is most commonly, although not in all cases exclusively, found. Each is followed by one or more code letters indicating the use or uses to which it is put by the Tsembaga; a key to these code letters follows Table 25.

Tentative bird identifications were made by Dr. Ralph Bulmer, and they may be found in Rappaport (1966). The native categories *ma* and *koi* have been glossed, for convenience, as marsupials and rats, respectively. These equivalences are only approximate, for the former may include giant rats while the latter may include small marsupials. It would be more accurate to define *ma* as wild mammals, excluding pigs and bats, over 4 or 5 inches high when standing on all fours, while *koi* are mammals under this size.

Table 21. Nondomesticated Resources Most Commonly Found in the Kamuŋga Geni Association (Primary Forest, 5,000′–7,000′)

Life form	Family	Genus & species	Native name	Uses*
		FLORA		
TREES	*Anacardiaceae*	*Semecarpus ?magnificans*	kuip	M
		?Pentaspadon or *Rhus* sp.	kariŋanč	TFn
	Cunoniaceae	*Caldcluvia* sp.	bokanč	B
	Euphorbiaceae	*Macaranga* sp.	konjenpai	Fl (smoked)
		Codiaeum variegatum	ger	R
	Fagaceae	*Quercus* sp.	nuŋ	B
	Guttiferae	*Garcinia* sp.	gun	R
		Garcinia sp.	tandapa	T
	Lauraceae	*Breilschmiedia* sp.	kom	RFl
		Cryptocarya sp.	kawit	RD
		Cryptocarya sp.	da	R
		Litsea sp.	dapai	RO
		?	krim	DTS

Table 21 (*continued*)

Life form	Family	Genus & species	Native name	Uses*
TREES *cont.*	*Lauraceae*	?	boko	R
		?	gumbiaŋ	R
	Leguminosae	*Albizzia* sp.	kanam	RT
	Loganiaceae	*Fagraea racemosa*	borumoi	S
	Melastomaceae	*Astronia* sp.	kukair	RS
	Moraceae	*Ficus calopilina*	muruŋ	XM
		Ficus dammaropsis	timnai	Ff
		Ficus trachypison	ringanč	BS
		Ficus sp.	gimbondum	XS
	Myrtaceae	*Eugenia* sp.	nonomba	RD
		Eugenia sp.	apeŋ	TD
		Decaspermum sp.	dam nene	B
	Palmae	*Calamus* sp.	kumbaka	T
		?	kumur	BSFs
	Pandanaceae	*Freycinetia* sp.	koraiŋga	T
		Pandanus sp.	buk	T
		Pandanus sp.	pima	BFf
		Pandanus sp.	taba	Ff
		Pandanus sp.	tumbama	BT
	Rubiaceae	?	ganč	R
		?	miñiŋgambo	M
	Rutaceae	?	konjup	RFl
	Theaceae	*Ternstroemia* sp.	tondomane	B
	Ulmaceae	?	dima	T
	Urticaceae	?	rama	Fl
	Winteraceae	*Bubbia* sp.	ruimam	D
	Unidentified		aimeŋga	MD
	“		ameŋgi	T
	“		air	T
	“		koro	B
	“		muŋr	RS
	“		gambo	TM
	“		nokopač	M
	“		nombon	X
	“		tiŋgia	T
SHRUBS & HERBS	*Chloranthaceae*	*Chloranthus* sp.	korap	RFl
			korap muŋa	
	Cyatheaceae	*Cyathea* sp.	noŋgam	D
	Gesneriaceae	*Cyrtandra* sp.	welenče	RFl
	Gramineae	*Bambusa* sp.	koa	TD
	Marantaceae	*Phrynium* sp.	miŋgin	BS
	Marattiaceae	*Angiopteris* sp.	doŋgai	R
	Palmae	?	mandiŋga	BS
	Piperaceae	*Piper* sp.	kere kere	R
		Piper sp.	čerap	S
	Pittosporaceae	*Pittosporum undulatum*	ambuŋgai	M
	Rutaceae	*Evodia anisodora*	tumbup	A
	Tiliaceae	*Microcos* sp.	ninkmai	Fl
	Urticaceae	*Elatostema* sp.	gañiŋgai	RM
	Zingiberaceae	*Alpinia* sp.	bañaŋgoi	RDS
		Alpinia sp.	puplaka	BS
	Unidentified		kopeŋga	XFl
	“		piŋgo	Fl
	“		morno	M
	“		gonbi	Fl
	“		puŋ	M
	“		tokmai	O
	“		kwipo	
	“		nink amp	RFl

Table 21 (*continued*)

Life form	Family	Genus & species	Native name	Uses*
VINES	*Cucurbitaceae*		gambroŋgin	T
			yibona	FlT
	Ericaceae	*Dimorphanthera* sp.	ayuk	BS
	Gesneriaceae	*Aeschynanthus* sp.	koramp andika	BOS
	Monimiaceae	*Palmeria* sp.	kep ndim	Fl
	Pandanaceae	*Freycinetia* sp.	kriŋa	XS
		Freycinetia sp.	kwiŋgaka	O
	Piperaceae	*Piper* sp.	morameka	RFl
	Urticaceae	*Elatostema* sp.	ap čembamai	M
		Pipturus sp.	deraka	RS
EPIPHYTES	*Cyatheaceae*	*Dawsonia* sp.	ka rawambo	O
	Lycopodiaceae	*Lycopodium* sp.	ap diŋgambe	R
	Orchidaceae	*Dendrobium* sp.	kanjkai	O
	Polypodiaceae	*Polypodium* sp.	kwiop	Fl

BIRDS (kabaŋ)
At least 19 native taxa, 18 of which are eaten, 12 of which are valuable for feathers. Most important are:

	Native name	Uses*
Greater Sickle-Bill	karanč	WF
Six Plumed Bird of Paradise	kiawoi	WF
Princess Stephanie Bird of Paradise	kombom	WF
Gardener Bower Bird	kombek	WF
Haryopsis sp.	binan	WF
Cassowary	kombri or yoŋge	WFT
King of Saxony Bird of Paradise	nomapunt	WF
Magnificent Bird of Paradise	pieŋmai	WF
Superb Bird of Paradise	yenandiok	WF

FUNGI (bai)
At least 8 native taxa, all unidentified, all used for food.

FAUNA

MARSUPIALS (ma): At least 16 native taxa, all unidentified, used for food, ornamentation, fiber, hides; some are wealth items.

RATS (koi): At least 6 native taxa, all used as food by women and children.

INSECT LARVAE (čuma): One native taxon, found in stumps, highly prized food.

* See notes to Table 25 for key to abbreviations.

Table 22. Nondomesticated Resources Most Commonly Found in the Tsembaga Wora Geni Association (Primary Forest, 2,200′–4,000′)

Life form	Family	Genus & species	Native name	Uses*
		FLORA		
TREES	*Anacardiaceae*	*Mangifera* sp.	wowi	Ff
		Spondius dulcis	aipan	R
	Araucariaceae	*Araucaria hunsteinii*	juk	T
	Casuarinaceae	*Casuarina papuana*	jimi	ASB
		Casuarina sp.	kepir	ASB
	Euphorbiacene	*Aleurites moluccana*	kaba	Fn
	Flacourtiaceae	*Pangium edule*	topia	R
	Leguminosae	*Erythrina* sp.	yaur	Fl
	Magnoliaceae	*Elmerrillia papuana*	banč	R
	Pandanaceae	*Pandanus* sp.	miyom	B
	Sapindaceae	*Planchonella* sp.	aŋa	A
	Vitaceae	*Leea* sp.	bebon	B
SHRUBS	*Amaryllidaceae*	*Crinum* sp.	yimane	T

FAUNA

BIRDS (kabaŋ)
At least 7 native taxa, all of which are eaten. The feathers of 4 are used as adornment, but only 2 are considered valuable. These are:

Genus & species	Native name	Uses*
Cacatua galerita	akaka	FW
Lesser Bird of Paradise	yambai	FW

MARSUPIALS (ma): At least 6 native taxa, all unidentified, used for food, hides, fiber, ornamentation, and wealth objects.

RATS (koi): At least 1 native taxon, used as food for women and children.

SNAKES (noma): At least 2 native taxa, used for food.

LIZARDS (tum): At least 1 native taxon, used for food and hide.

* See notes to Table 25 for key to abbreviations.

Table 23. Nondomesticated Resources Most Commonly Found in Tsembaga Ringopwai (Secondary Growth) Associations

Life form	Family	Genus & species	Native name	Uses*
		FLORA		
TREES	*Cyatheaceae*	*Cyathea angiensis*	yimunt	Fl
		Cyathea, new species	kaŋgup	Fl
	Ebenaceae	*Diospyros* sp.	wonum	M
	Euphorbiaceae	*Breynia* sp.	non manč	R
		Phyllanthus sp.	dikambo	B
			jimbonk	RM
	Melastomaceae	*Melastoma malabathricum*	wop kai	X
	Moraceae	*Ficus* sp.	danje	T
	Palmae	*Licuala* sp.	moropmai	T
	Rhamnaceae	*Alphitonia iacana*	pokai	BM
	Sapindaceae	*Dodonaea viscosa*	gra	B

Table 23 (*continued*)

Life form	Family	Genus & species	Native name	Uses*
TREES *cont.*		*Mischocodon* sp.	birpi	B
	Saurauiaceae	*Saurauia* sp.	rokunt	B
	Solanaceae	*Lycianthes* sp.	kapaŋ	R
		Solanum sp.	gon	RM
	Ulmaceae	*Gironniera* sp.	penta	B
	Urticaceae	*Missiessya* sp.	yamo	FlSM
	Verbenaceae	*Geunsia farinosa Bl.*	gonwant	M
	Unidentified		yent	B
	"		raŋgan	T
	"		kop	X
	"		mar mar	B
HERBS & SHRUBS	*Acanthaceae*	*Rungia klossi*	tok mai čeŋmba	Fl
	Araceae	*Aglaonema* sp.	močam	RD
		Alocasia sp.	gump	S
	Commelinaceae	*Commelina* sp.	komerik	Fl
	Compositae	*Bidens pilosa*	womemuk	MO
		Blumea sp.	gumbandi	M
	Cyatheaceae	*Cyathea rubiginosa*	kabaŋ bep	Fl
	Gramineae	*Bambusa forbesii*	kinjen	B
		Coix lachrymajobi	koŋgun	OA
		Imperata cylindrica	korndo	B
		Miscanthus floridulus	ripa	T
		Phragmites karka	yamboč	M
		Setaria palmaefolia	korami	Fs
	Orchidaceae	*?Calanthe* sp.	korndo tiokum	SR
		Spathoglottis sp.	korndo tiokum	SR
	Piperaceae	*Piper* sp.	yikun	RD
	Polypodiaceae	*Dennstaedtia* sp.	terai	Fl
	Urticaceae	sp.	nent	M
	Zingiberaceae	*Costus* sp.	monomp	B
		Riedelia sp.	yenjim	M
		Cyclosorus sp.	aruk	Fl
		Diplazium sp.	raŋgilopa	Fl
	Unidentified		rum rena	Fl
VINES	*Asclepiadaceae*	*Hoya* sp.	koiwundo	SB
	Gesneriaceae	*Aeschynanthus* sp.	yimbuŋk	SB
	Liliaceae	*Smilax* sp.	gum biogun	T
	Melastomaceae	*Medinella* sp.	aikumbindi	B
	Moraceae	*Ficus* sp.	mopakai	SB
	Pandanaceae	*Freycenetia* sp.	kwiŋgaka	O
	Passifloraceae	*Adenis* sp.	akar	SB

FUNGI (bai): At least 7 native taxa used as food.

FAUNA

SNAKES (noma): At least 4 native taxa, all used for food, one for hide (python).

INSECTS (baŋ): At least 8 native taxa, 7 used for food, 1 for medicine, 1 for ornament.

* See notes to Table 25 for key to abbreviations.

Table 24. Nondomesticated Resources Most Commonly Found in Streams and on Stream Banks in Tsembaga Territory

Life form	Family	Genus & species	Native name	Uses*
		FLORA		
TREES	*Myrtaceae*	*Cleistanthus* sp.	timbi	R
	Unidentified		pima	R
			kumbent	
BUSHES, SHRUBS, HERBS	*Moraceae*	*Ficus andenosperma*	anjai	Fl
	Myrtaceae	*Eugenia* sp.	druo	R
	Urticaceae	*Laportea* sp.	čenaŋ	Fl
			gumiŋga	

FAUNA

MARSUPIALS (ma): At least 3 native taxa, used for food.

BIRDS (kabaŋ): At least 3 native taxa used for food.

FROGS (kamp): Many taxa, used for food by women and children.

CRABS (korapa): One native taxon, used for food.

FISH: Two native taxa, *kobe* (eel), and *tuoi* (catfish), both eaten; *kobe* also used in rituals.

* See notes to Table 25 for key to abbreviations.

Table 25. Nondomesticated Resources Commonly Found in Two or More Associations in Tsembaga Territory

Life form	Family	Genus & species	Native name	Uses*
		FLORA		
TREES	*Araliaceae*	*Boerlagiodendron* sp.	aimam	B
	Dilleniaceae	*Dillenia* sp.	munduka	SB
	Euphorbiaceae	*Homolanthus* sp.	tup kalom	X
	Leguminosae	*Desmodium sequax*	korai	O
			jnjie	
	Meliaceae	*Chisocheton* sp.	birpi	Fn
	Moraceae	*Ficus puncens*	kobenum	Flf
		Ficus wassa	beka	XFlf
	Myrtaceae	*Descapermum necrophyllus*	jinjimbint	R
	Thymelaceaceae	*Phalerium nisidai*	pukna	X
	Urticaceae	*Maoutia* sp.	noŋgamba	X
		Oreocnida sp.	rumem	B
	Unidentified		membra	B
	"		dukumpna	Flf
	"		pia	B
	"		punt	B
HERBS & SHRUBS	*Gramineae*	*Bambusa* sp.	waia	BT
	Lycopodiaceae	*Dimorphanthera* sp.	koropi mai	M
	Zingiberaceae	*Alpina* sp.	anjomar	S
			gaunai	M
	Unidentified		mi muŋga	M

Table 25 (*continued*)

Life form	Family	Genus & species	Native name	Uses*
VINES	*Convolvulaceae*	*Lepistemon urceolatum*	apop	SB
	Cucurbitaceae	?*Trichosanthes* sp.	jen	Ff
	Flagellariaceae	*Flagellaria indica*	goŋ	SB
	Polypodaceae	*Dicranopteris pinearis*	mombo	B
	Zingiberaceae	?*Alpinia* sp.	čawaka	SB
	Unidentified		mundunt	SB

FAUNA

BIRDS (kabaŋ): At least 40 native taxa, used for food, ornamentation, and wealth objects.

MARSUPIALS (ma): At least 3 native taxa, used for food, fiber, ornamentation, and wealth objects.

RATS (koi): At least 1 taxon, used as food by women and children.

WILD PIG (konj): 1 taxon, used as food.

SNAKES (noma): At least 2 taxa used for food.

* Uses of Plants and Animals:

A. Agronomic—Used to protect other plants, or to improve their yields; plants that are sometimes planted to improve the associations that develop on abandoned gardens; and indicator plants (plants whose achievement of a particular stage of maturation or a particular stage in a reproduction cycle serves as a signal to undertake one or another subsistence activity).

B. Building Materials—Used in the construction of quasi-permanent structures, usually houses and fences.

D. Dress—Used without processing as part of the costume; usually leaves used as coverings for buttocks.

F. Food—Only materials customarily ingested by humans. It does not include materials habitually consumed by livestock and wild animals but not human beings. In the case of plant foods, the part of the plant used is designated by a small letter: (f) fruit; (l) leaf; (n) nut; (r) root; (s) stem.

M. Medicinal, Hygienic, Cosmetic—Used in any procedure "magical," "religious," or empirical, the aim of which is to cure an existing physical disability. Also included are any materials expended in procedures that produce or maintain personal cleanliness, and materials which are used as cosmetics.

O. Ornamentation—Worn on the person as adornment, but do not, like cosmetics, lose their physical properties after a single wearing.

R. Ritual—Used in rituals with the exception of those included in category M.

S. Supplies—Used once, then discarded, e.g., leaves for temporary packages, vines to tie firewood, leaves used in earth ovens. Items included in categories B, D, F, and M are excepted.

T. Technological—Used to make tools, weapons, or containers that are not discarded after being used once.

W. Wealth Objects—May be exchanged for steel tools, shell objects and, in former times, stone axes (most importantly feathers and furs).

X. Important for their fibers or dyestuffs, which are used in the manufacture of bark cloth, net bags, loin cloths, etc.

APPENDIX 9

Diet

MEASUREMENT OF THE NUTRITIVE VALUE OF TSEMBAGA INTAKE DERIVED FROM PLANTS

Fruits and vegetables comprise approximately 99 percent, by weight, of the usual daily Tsembaga intake.

Total Intake, by Weight, of the Tomegai

All vegetable foodstuffs brought home to the four hearths of the Tomegai clan were weighed daily, by named variety, from February 14, 1963, to December 14, 1963. For the purpose of estimating intake the figures from March 11 to November 8, a period during which both the pig and human populations were relatively stable, form the basis of the estimate.

From the gross weights brought home, two deductions were made. First, the harvester was asked to set aside the ration for pigs. The weight of that ration was subtracted from the gross weight, leaving the weight available for human consumption. Second, a factor for waste in preparation was subtracted from the remaining amount. The values assigned to waste are in the case of each item based upon tests in the field. The figures remaining after the two deductions indicate total weights of edible portions brought to the houses.

To these figures are added values for the consumption of vegetable foods away from the house. These values are based upon observations of people eating on the paths, in the bush, and in the gardens. The resulting figures are taken to be total weight of edible portion available for human consumption.

The procedures for obtaining this final figure may be summarized in the following equation:

(Gross weight brought home + estimated weight consumed away from home) − (pig ration + waste) = total weight of edible portion available for consumption by humans.

These figures are tabulated in columns 1–5 of Table 26.

It may be that some of the difficulties in deriving quantities of edible portion available for human consumption are apparent. In common with the attempt to estimate garden yields, there is the problem of estimating the weights of foodstuffs that were consumed without having first been brought to the houses. Values were assigned on the basis of observations made without the benefit of scales. The quantities consumed away from the houses are not large in any event, and it is therefore not likely that estimation resulted in an error of more than one or two percent except in the cases of sugar cane and cucumbers, where error could have been greater.

The values assigned to the pig ration reflect the weight of foodstuffs explicitly set aside for them. However, women were frequently observed feeding the pigs tidbits from the portion set aside for humans. No deductions were made from the human portion for these morsels. The daily human ration is therefore represented as slightly higher, in comparison to the ration set aside for pigs, than it actually is.

The percentage values assigned to waste were based upon a limited number of tests in each case, and the amount of waste varies with the age, freshness, size, etc., of specimens, even within particular varieties of the same species. The values I assigned on the basis of tests do, however, approximate standard values found in the literature.

The Value of Food Ingested

Nutritive values derived from the literature were assigned to each item. Wherever possible, values used by other New Guinea workers were adopted to facilitate comparability. In most cases these workers adopted values derived from samples taken from the general area in which they worked. Other things being equal, it seemed methodologically preferable to accept values derived from the analyses of New Guinea specimens rather than values taken from specimens grown elsewhere. In some instances the values used may be high. The caloric value for sweet potato, for instance, derived from Hipsley and Clements (1947), is 681 per pound of edible portion. Other authorities record considerably lower figures. It is upon the basis of Hipsley's and Clements' values, however, that the Chimbu, Busama, Kaiapat, Patep, and Kavataria diets were evaluated, and the adoption of the same values, so far as possible, facili-

Table 26. Total Intake of the Tomegai Clan

Crop	1 Total weighed, in lbs.	2 Ration for pigs, in lbs.	3 Available for humans, in lbs.	4 Edible portion factors, in %	5 Edible portions for humans, in lbs.	6 Calories/lb. edible portions	7 Authority*	8 Calories, human intake	9 % Protein edible portion	10 Authority*
Roots										
Sweet potato	9,944.50	5,554.22	4,390.28	80	3,512.23	681	3	2,391,828.63	.9–1.7	5
Xanthosoma	1,505.50	14.50	1,491.00	80	1,192.80	658	3	784,862.40	1.4–1.9	5
Colocasia	3,834.75		3,834.75	85	3,159.54	658	3	2,078,977.32	1.4–1.9	5
Manioc	1,349.25	1,106.39	242.86	80	194.29	595	3	115,602.55	.7–1.2	5
Yams	1,942.50		1,942.50	85	1,641.60	486	3	652,017.68	1.9–2.0	5
Total roots								6,023,288.58		
Trees										
Banana	1,682.50	54.90	1,627.60	70	1,139.32	427	3	486,489.64	1.1	3
Artocarpus	105.50		105.50	50	52.75	295	1	15,561.25	1.0	1
Marita	861.50		861.50	43	370.45	762	7	282,282.90	3.7	7
Total trees								784,333.79		
Misc. Garden										
Corn	8.50		8.50	29	2.47	463	6	1,143.61	3.7	6
Peas and beans	4.75		4.75	95	4.51	440	3	1,984.40	8.1	3
Pumpkin	234.25	33.26	200.99	68	136.67	200	3	27,334.00	1.5	3
Gourd	13.75		13.75	68	9.35	154	1	1,439.90	1.3	1
Total misc. garden								31,901.91		
Leaves										
Hibiscus	1,373.75		1,373.75	95	1,305.06	136	3	177,488.16	5.7	2
Rungia klossi	316.75		316.75	95	300.91	136	3	40,923.76	3.8	2
Sweet potato leaves	138.25		138.25	95	131.34	218	5	28,632.12	3.6	5
Assorted others	244.50		244.50	95	232.28	136	3	31,590.08	3.8	2
Total leaves								278,634.12		
Grasses and Inflorescences										
Pitpit-dičiŋ variety	1,142.00		1,142.00	40	456.80	104	4	47,507.20	4.1	2
Pitpit-dičiŋ, trimmed	124.75		124.75	95	118.51	104	4	12,325.04	4.1	2
Pitpit, all other var.	816.25		816.25	60	489.75	104	4	50,854.00	4.1	2
Pitpit, all other var. trimmed	61.75		61.75	95	58.66	104	4	6,100.64	4.1	2
Setaria	692.75		692.75	17	117.77	101	8	11,994.75	1.4	8
Setaria, trimmed	61.75		61.75	95	58.66	101	8	5,924.90	1.4	8
Total grasses and inflor.								134,706.53		
Refreshers										
Sugar cane	3,368.00	249.50	3,118.50	30	935.55	263	3	246,049.65	.4	3
Cucumbers	216.50	10.80	205.70	95	195.42	50	3	9,771.00	.6	3
Total refreshers								255,820.65		
Totals, all sources								7,508,685.58		

* Authorities:

(1) FAO Nutritional Study No. 11.
(2) Hamilton, 1955.
(3) Hipsley and Clements, 1947.
(4) Hitchcock, N. 1962.
(5) Massal and Barrau, 1956.
(6) Osmund and Wilson.
(7) Peters, 1958.
(8) Wenkam, K., Department of nutrition, University of Hawaii, personal communication.

(16 Persons, 13 Pigs; March 11–November 8, 1963)

11	12	13	14	15	16	17	18	19	20	21	22
Protein ingested by humans, minimum lbs.	Protein ingested by humans, maximum lbs.	Factors for additional intake, in %	Total estimated caloric intake	Total estimated Protein intake, minimum lbs.	Total estimated Protein intake, maximum lbs.	Fat, in %	Authority *	Total fat intake, in lbs.	Calcium, mgm./lb.	Authority *	Total calcium intake, mgm.
31.61	59.71	5	2,511,420.06	33.19	62.70	.3	5	10.53	68	3	2,188.16
16.70	22.66	5	824,105.52	17.54	23.79	.2	5	2.39	177	3	2,029.84
44.23	60.03	5	2,182,926.19	46.44	63.03	.2	5	6.32	177	3	5,591.43
1.36	2.33	5	121,382.68	1.43	2.45	.3	5	.58	114	3	221.16
25.49	26.83	5	684,618.56	26.76	28.17				45	5	738.45
12.53	12.53	20	583,787.57	15.04	15.04	.5	1	5.70	36	3	410.04
.53	.53		15,561.25	.53	.53	.6	1	.32	168	1	87.36
13.71	13.71	25	289,339.97	17.14	17.14	14.0	3	5.19			
.91	.91	5	1,150.79	.96	.96	4.3	1	1.06	32	1	.64
.37	.37		1,984.40	.36	.36	3.9	3	1.76	136	6	5.44
2.05	2.05		27,334.00	2.05	2.05				91	3	23.76
.11	.11		1,439.90	.12	.12	.2	1	.02	54	1	4.86
74.38	74.38	2	181,037.92	75.87	75.87	.3	4	1.92	1,685	4	21,989.25
11.43	11.43	2	41,542.24	11.66	11.66	.3	4	.90	1,685	4	5,055.00
4.73	4.73	2	29,204.76	4.82	4.82	.7	5	.92	340	5	445.40
8.83	8.83	2	31,921.89	9.01	9.01	.3	4	.70	1,685	4	3,909.20
18.73	18.73	2	48,457.34	19.10	19.10	.2	3	.91	95	3	433.20
4.86	4.86	2	12,471.54	4.96	4.96	.2	3	.14	95	3	112.10
20.08	20.08	2	51,871.08	20.48	20.48	.2	3	.98	95	3	464.55
2.41	2.41	2	6,222.65	2.46	2.46	.2	3	.12	95	3	51.50
1.65	1.65	2	7,808.15	1.68	1.68	.6	8	.71	95	8	111.15
.72	.72	2	3,889.16	.73	.73	.6	8	.35	95	8	51.50
3.74	3.74	200	738,148.95	11.22	11.22				45	3	420.75
1.17	1.17	50	14,656.50	1.36	1.36				104		202.80
302.33	354.50		8,412,483.06	324.91	379.69			41.52			44,547.54

tates comparisons within Australian New Guinea. It may be, however, that these values should be adjusted in any comparison with nutritional data collected in other parts of the world.

Nutritional values are summarized in columns 6, 9, 17, 20 of Table 26. The sources from whom the figures were adopted appear in adjacent columns.

Nutritional Values of Individual Intakes

The sample for which quantitative consumption data were collected consisted of sixteen persons for a period of 246 days. Not all sixteen were present every day, and on some occasions visitors were fed; therefore a record was kept of who ate at each hearth every day. The persons who were fed fell into a number of different age and sex categories, each of which had a different intake. The problem of assigning values to the intake of each age-sex category theoretically can be solved simply: one merely measures the quantities served to persons of each category. Among the Maring such a procedure was not practical. It would have required first that the observer be on hand with a portion scale while the food was being consumed, and left-overs may be eaten late at night or early in the morning, when tubers may also be cooked for later consumption as between-meal snacks. Unless the observer takes up more or less permanent residence next to the larder of a single household, some of this consumption is bound to escape his attention.

As an alternative, ratios for the quantity of intake of persons in the various age and sex categories were devised. These were based upon suggested caloric allowances published by Venkatachalam (1962:10), following Langley (1947:134). The daily caloric allowances listed here are proposed by Venkatachalam and Langley for the categories included in the sample:

Adult males	2,500
Adult females	2,100
Adolescent females	2,050
Children 5–10 years	1,300
Children 3–5 years	1,200
Children 1 year	850

The ratio of these values was assumed to reflect the comparative ingestion of persons in the various age-sex categories. An adult female, for instance, was assumed to eat 21/25 as much as an adult male. A child five to ten years of age was assumed to eat 13/25 as much as an adult male, and so on. The problems of such a procedure are clearly understood. It is no doubt the case that actual practice deviates from ideal apportionment patterns. This procedure is adopted as the lesser of the possible evils: I believe greater error would have resulted from an attempt to estimate the comparative sizes of actual portions.

The suggested caloric values were divided by 100; the resultant figures may be referred to as "trophic units." An adult male represents twenty-five trophic units, an adult female twenty-one, etc. These were then multiplied by the number of consumer days in each category, yielding the number of trophic unit days in each, and a total of trophic unit days for the entire sample was taken. The total values of the several nutrients were then divided by the total number of trophic unit days to arrive at a trophic unit day value for each nutrient. From this, the daily intake of each nutrient for each category of persons was derived.

COMPARISON OF TSEMBAGA AND OTHER NEW GUINEA DIETS

Table 27 compares the Tsembaga diet to that of five other New Guinea groups. The figures published for Busuma, Kaiapit, Patep, and Kavataria

Table 27. Comparative Values of Six New Guinea Diets

Place	Source	Daily calories	Daily protein, in grams			Daily calcium, in grams	Total daily intake, in grams
			Veg.	Animal	Total		
Busama	Hipsley and Clements, 1947	1,223	14.4	4.7	19.1	.5	794
Kaiapit	Hipsley and Clements, 1947	1,609	21.7	3.1	24.8	.6	1,013
Patep	Hipsley and Clements, 1947	1,904	22.3	2.1	24.4	.6	1,387
Kavataria	Hipsley and Clements, 1947	1,600	22.4	18.4	41.3	.3	1,256
Chimbu	Venkatachalam, 1962	1,930	20.8	?	20.8 +		1,627
Tsembaga, minimum			34.7	?	34.7 +		
Tsembaga, maximum		2,015	46.8	?	46.8 +	1.2	2,287

are total per capita figures. No breakdown by age or sex categories is provided. It was therefore necessary to determine overall per capita figures for the intake of various nutrients by Tsembaga. The usefulness of such figures is limited to purposes of comparison. For purposes of evaluation, separate figures by age and sex category are preferable.

The information summarized in Table 27 suggests the marked superiority of the Tsembaga diet over those of the other groups. The differences between the Tsembaga and several of the other groups are so large, however, that they are suspect. The size of the disparities becomes increasingly apparent when gross intake is compared. Langley reports Busama intake to be 794 grams per day (1947:112–15). Tsembaga intake was almost three times as great. Langley states that there was a food shortage at Busama at the time of the survey, but no mention is made of a food shortage at Kaiapit, where the reported daily intake was only 1,013 grams, less than half that of the Tsembaga.

The methods employed by the New Guinea Food Survey expedition in 1947 are not made fully explicit, so it is impossible to judge whether different procedures led to different results. My belief is, nevertheless, that figures for Busama, Kaiapit, Patep, and Kavataria, which were derived from a limited number of visits to houses at mealtimes, do not adequately represent per capita consumption in those communities. With the exception of Patep, they seem much too low.

The Chimbu figure lends support to the Tsembaga figure. Much of the difference in the weights of Tsembaga and Chimbu intakes lies in the availability to the Tsembaga of greater amounts of nonstarch vegetables. It is in this difference that the qualitative as well as quantitative superiority of Tsembaga diet over Chimbu lies. Apparently absent from the diet of the latter are some of the leaves that constitute the most important source of both protein and calcium for the Tsembaga. It is probably altitudinal limitations that deprive the Chimbu of these valuable greens.

APPENDIX 10

Carrying Capacity

CARRYING CAPACITY FORMULA

Carneiro's formula for carrying capacity is the one used here.

$$(1) \qquad P = \frac{\frac{T}{R+Y} \times Y}{A}$$

Where:

P = The population which may be supported.
T = Total arable land.
R = Length of fallow in years.
Y = Length of cropping period in years.
A = The area of cultivated land required to provide the "average individual" with the amount of food he ordinarily derives from cultivated plants per year.

THE VALUES OF VARIABLES T AND R

The total area of arable land is not easy to determine in terrain as rugged as that occupied by the Tsembaga. The difference between orthographic and surface areas is substantial, due to both general and localized sloping. My calculation is that Tsembaga territory comprises, in total, 2,033 acres by orthographic measurements made on an aerial photograph. The total territory may be divided into a number of areas in terms of altitude, past use, vegetation cover, and agricultural potential. These discriminations are summarized in Table 28.

The 100 ± acres estimated to be arable in the Jimi Valley and the 200 ± acres estimated to be arable in the high altitude virgin forest in the Simbai Valley were not measured. These figures represent estimates, based upon walks through the areas, of the comparative size of areas un-

Table 28. Tsembaga Territory, Arable and Nonarable Land

Area	Total acreage	Arable	Nonarable
High-altitude high forest, Jimi Valley	343	100 ±	243
High-altitude high forest, Simbai Valley	602	200 ±	402
Secondary forest (gardened areas), Simbai Valley	1,019	864	150
Low-altitude forest remnants	28		28
Grassland	41		41
Totals	2,033	1,164	864

der high forest and nonarable moss forest. The border between these areas is indistinct. The area under high forest in the Simbai Valley seems never to have been gardened, and no gardening has been undertaken by Tsembaga in the Jimi Valley area for many years.

The figure in Table 28 for grassland includes only the one extensive area noted in Chapter 3. Allowance has been made for other, smaller grassy areas in the figure for nonarable land within the secondary forest. Similarly, the figure for low-altitude forest includes only areas of 3 to 4 acres or more. Smaller patches, like the smaller grassy areas, are included in the figure for nonarable land within the secondary forest. This latter figure is an estimate that allows for stream beds, rocky areas, gullies, and slopes over 43°, as well as for the small grassy areas and small patches of forest. The figure of 150 acres is probably conservative.

The areas tabulated above are orthographic areas. The terrain, however, is very rugged. The general slope in the areas that can be gardened is about 20°. This increases surface over planimetric area by a factor of 16.15:15. The intricate dissection of the terrain by streams and spurs projecting from the mountain wall further complicates the surface and increases its area, but there is no means for judging by how much. It will be assumed here that slopes in other directions approximate the general slope. Surface area will thus be taken as $16.15^2:15^2$, or 116.48 percent of orthographic area. Arable surface is summarized in Table 29.

Table 29. Arable Land Area Corrected for Slope, Tsembaga Territory

Location and association	Acreage
Simbai Valley secondary forest and gardens	1,002
Simbai Valley high forest	235
Jimi Valley high forest	116
Economic population density, total arable area:	97 p.s.m.
Economic population density, secondary forest and gardens:	124 p.s.m.

It is not the case, however, that all areas are equally productive. In an earlier section it was shown that the sweet potato gardens that are planted at higher altitudes in the secondary forest produce only 92.6 percent as many calories (4,418,215) in the twelve months of maximal production as do the taro-yam gardens planted at lower altitudes. An adjustment must be made for this difference. Again, however, a distinct boundary between the two zones cannot be drawn, for the region between 4,000 and 4,400 feet altitude is transitional. The lower zone will be said to consist of 650 acres and the upper, slightly less productive, areas 352.

It is reasonable to expect that gardens that might be cut in the high-altitude virgin forest, would, in the long run, show even lower productivity than those in the upper portion of the secondary forest. It will be assumed here that the diminishment in productivity would be of an order similar to that between the lower two zones. Arable land in the virgin forest will be regarded as 92.6 percent as productive as land in the upper portion of the secondary forest. Differences in lengths of fallow, as well as differences in yields, must also be considered. In an earlier section a value of fifteen years was adopted for the lower secondary forest zone and twenty-five years for the upper secondary forest zone. It is reasonable to expect that fallows on gardens cut in the virgin forest between 5,200 and 6,000 feet would be even longer. A figure of thirty-five years will be assumed here.

The three land classes that have been distinguished, differing both in productivity and length of fallow, are summarized in Table 30. These

Table 30. Tsembaga Arable Land Classes

Class	Acres	Productivity	Frequency of use	
I	650	1.000	15 years	1.000
II	352	.926	25 years	.600
III	351	.857	35 years	.430

productivity and frequency of use factors permit the areas included in the three classes to be reduced to equivalent figures. Areas in Classes II and III, that is, may be expressed in terms of equivalent amounts of Class I land.

THE VALUE OF VARIABLE A

The value for A, the amount of land required for the "average individual," is a function of the age-sex composition of the particular population under study. The figure used here is based upon the age-sex composition of the Tsembaga in 1963. If there were changes in the proportions of persons in the various age-sex categories, each of which have different trophic requirements, there would be a change in the value of A.

Values for Variable A, Pig Population at Minimum

Actual measurements, based upon chain and compass mapping, are available for total acreage in production when the pig population was at a maximum. The areas of the gardens of the Tomegai clan are also available for 1963, when the herd was reduced to a minimum. These and other figures permit the determination of variable A for the Tsembaga when their pig herd was at a minimum in three ways. All three calculations will be made here as a check upon each other.

(A) The total annual caloric requirements of the entire population, with the pig herd at a minimum, may be divided by the annual per-acre yields of taro-yam gardens. The production of taro-yam gardens, rather than sugar-sweet potato gardens, may be taken because, as it has already been mentioned, virtually no separate sugar-sweet potato gardens are planted when the pig herd is at its minimum. Actual weighings have demonstrated that 85 percent of the total caloric yield of the taro-yam gardens is harvested during one twelve-month period. It is this figure, 4,418,215 calories, which is, accordingly, used in this calculation.

Total Tsembaga consumption is presented in Table 31. The number of

Table 31. Total Caloric Requirements, All Tsembaga

Age	Sex	Trophic units	Daily per capita intake (cal.)	Number of persons in category	Total daily consumption of category (cal.)
Over 50	male	20	2,060	8	16,480
	female	16	1,648	12	19,776
21–49	male	25	2,575	49	126,175
	female	21	2,163	40	86,520
15–20	male	29	2,987	11	32,857
	female	20.5	2,112	7	14,784
10–15	male	22	2,266	15	33,990
	female	21	2,163	8	17,304
5–10	male-female	15	1,339	25	33,475
0–5	male-female	10	1,030	29	29,870

Total daily consumption: 411,231 cal.
Total annual consumption: 150,089,315 cal.
Total trophic units: 4,042

trophic units represented by categories not included within the Tomegai clan, from whom all consumption data were gathered, are derived from Langley (1947:134).

The amount of acreage (Gs) that must be put into production annually to fulfill Tsembaga trophic requirements may be solved for by the following simple equation:

$$(2) \qquad Gs = \frac{Ct - Co}{Ca}$$

Where:

Ct = Total caloric requirements, Tsembaga.
Co = Calories available from old gardens (15% of total).
Ca = Caloric yield per acre, taro-yam gardens, 24–76 weeks after planting.

$$Gs = \frac{150{,}089{,}315 - 22{,}513{,}389}{4{,}418{,}215}$$

A value may now be assigned variable A through solution of the following equation:

$$(3) \qquad A = \frac{Gs}{P}$$

Where:

P = Total Tsembaga population (204 persons).

$$A = \frac{28.87}{204}$$

$A = .142$ acres.

(B) The acreage per "trophic unit" placed in production by the Tomegai clan in 1963 may be calculated from the total area they placed in production in 1963 by solution of the following equation:

$$(4) \qquad Ua = \frac{Gn}{U}$$

Where:

Ua = Acreage per trophic unit.
Gn = Acreage put into production by the Tomegai, 1963 (3.07 acres).
U = Total trophic units, Tomegai clan (313).

$$Ua = \frac{3.07}{313}$$

$Ua = .0098$ acres

If the acreage per trophic unit is then multiplied by the total trophic units of all the Tsembaga, a figure for minimal required acreage is obtained.

(5) $$4042 \times .0098 = 39.61 \text{ acres}$$

$$A = \frac{39.61}{204}$$

$$A = .194 \text{ acres}$$

(C) Another calculation may be made by extending a comparison of the acreages placed in production by the Tomegai in 1962 and 1963 to the acreage of all of the Tsembaga.

The following areas are known by measurement:

Gl = Total acreage put into production, 1962.
Gt = Acreage put into production by Tomegai, 1962.
Gn = Acreage put into production by Tomegai, 1963.

To solve for total acreage put into production in 1963 (Gs), the following ratio may be used:

(6) $$Gl{:}Gs = Gt{:}Gn$$

Correction must be made for the differences in the ratios of sugar-sweet potato gardens to taro-yam gardens in the two years. This may be accomplished by using the factor that converts Class II to Class I to land.

Tomegai gardens, 1962 (actual measurement)	
Area in acres, sugar-sweet potato gardens:	
2.56 corrected:	2.37
Area in acres, taro-yam gardens	2.27
Total acreage corrected:	4.64

Tomegai gardens, 1963 (actual measurement)	
Area in acres, sugar-sweet potato gardens:	
.19 corrected:	.18
Area in acres, taro-yam gardens:	2.89
Total acreage corrected:	3.07

All Tsembaga gardens, 1962 (actual measurement)	
Area in acres, sugar-sweet potato gardens:	
19.34 corrected:	17.91
Area in acres, taro-yam gardens:	27.84
Total corrected:	45.75

Values are now available for solution of ratio (6).

(6) $$45.75\!:\!Gs = 4.64\!:\!3.07$$

$$Gs = 30.27$$

(7) $$A = \frac{30.27}{204}$$

$$A = .148 \text{ acres}$$

The discrepancy between calculations 1 and 2 is .052 acres per person, in aggregate 10.74 acres, or approximately 27%. Discrepancies of this magnitude are, perhaps, to be expected when values rest in part upon measurements undertaken in terrain as broken as that occupied by the Tsembaga and upon the extension of measurements of intake, themselves subject to some imprecision. It is probable, however, that this discrepancy has its basis not so much in errors in measurement as in actual differences in gardening practice. Calculation C bears out this interpretation. It will be noted that the Tomegai clan comprises 313 "trophic units," 7.77% of the total of 4,042 "trophic units" for the entire Tsembaga. In 1962, however, Tomegai gardens comprised 4.55 acres (corrected) or 9.94% of the total of 45.75 acres (corrected) put into production by all Tsembaga. The discrepancy between 7.77% and 9.94%, approximately 22%, corresponds closely to the discrepancy between calculations B and C, and may cancel it. It may be, that is, that the Tomegai regularly put larger-than-average amounts of land into cultivation, or at least did so in 1963 as well as 1962. This would not affect the reliability of derivation C, nor would the reliability of derivation A be affected unless it could be shown that the Tomegai eat more than other Tsembaga, and that there has therefore been an error made in extending Tomegai trophic requirements to the entire local population. There is no reason to believe this. Tomegai adult males are slightly below (approximately 3 kg) the average weight of all Tsembaga adult males, and the average weight of Tomegai adult females falls almost exactly on the average for the Tsembaga adult females as a whole. Moreover, the Tomegai do not seem to be either more or less active than other Tsembaga.

Their greater-than-average acreage in cultivation in 1962 cannot be explained, secondly, by the possession of a greater-than-average number of pigs by the Tomegai, since this was not the case. The ratio of pigs (of 120- to 150-pound size) to people for all the Tsembaga in 1962 was .83:1. The ratio of Tomegai pigs to Tomegai was .81:1.

Another explanation might have to do with land quality. There is no reason to believe, however, and considerable reason to reject, the notion that Tomegai gardens were on poorer-than-average land in 1962 or, as a rule, at other times.

Two explanations may be suggested. First, Tomegai male gardeners included six men making all their gardens and one man making half his gardens with the four Tomegai women and the one Tomegai adolescent girl (whose gardening activities may be counted as half that of an adult woman). This ratio of 6.5 male gardeners to 4.5 female gardeners exceeds the average, which is approximately 6.7:5.7. It is likely that with more males engaged in clearing, more acreage will be put into production.

Second, the discrepancy may be random. It is not likely that any group of ten or eleven gardeners will put into production the same amount of acreage as any other group of similar size, even within the same local population. The Tomegai may be particularly industrious (although it didn't seem so) or it may be that the measurements were taken in years when their acreage under production exceeded the average because of such considerations as the configuration of the land on the sites they were gardening. I believe the latter to be most likely.

It is probable that errors in measurement, sampling error, and differences in actual practice have all contributed to the discrepancy between calculations 1 and 2. The indication is, however, that a value of .194 for variable A is too high, and that more realistic values are derived in the two other calculations. Both extremes, however, will be used in further calculations.

Value for Variable A, Pig Population at Maximum

Since actual measurements were made of the total area put into production when the pig population was at its maximum, no estimate need be made. The amount, 47.18 acres, has already been introduced in previous calculations, where an adjustment to reduce it to its Class I equivalent, 45.75 acres, was made.

The formula solving for variable A may now be applied for acreage per capita under production when the pig population is at maximum.

$$A = \frac{45.75}{204} \tag{8}$$

$$A = .224 \text{ acres}$$

ESTIMATION OF CARRYING CAPACITY FOR HUMANS

Values for all variables are now provided, enabling us to solve for the maximum number of human beings that could be supported on Tsembaga territory by using formula (1).

Carrying Capacity of Secondary Forest Areas

Separate calculations will be made for the maximum number of people who could be supported on areas either under cultivation or under secondary forest at the time of fieldwork when (1) the pig population is at minimum, (defined as the ratio pigs : people = .29:1, pigs averaging 60 to 75 pounds) and (2) when the pig population is at maximum, (defined as the ratio pigs : people = .83:1, pigs averaging 120 to 150 pounds).

CARRYING CAPACITY, PIG POPULATION AT MINIMUM

Three values for variable A were derived; calculations will be made using all three.

(1)
$$P = \frac{\frac{T}{R + Y} \times Y}{A}$$

$$P = \frac{\frac{T \text{ class I} + (T \text{ class II} \times .926 \times .6)}{(R + Y)} \times Y}{A}$$

a) A = .142 acres

$$P = \frac{\frac{650 + (352 \times .926 \times .6)}{14 + 1} \times 1}{.142}$$

$$P = \frac{\frac{846}{15}}{.142}$$

$$P = \frac{56.40}{.142}$$

P = 397 persons

b) A = .148 acres

$$P = \frac{56.40}{.148}$$

P = 383 persons

c) A = .194 acres

$$P = \frac{56.40}{.194}$$

$$P = 290 \text{ persons}$$

CARRYING CAPACITY, PIG POPULATION AT MAXIMUM

Only one value for variable A will be used, since it was obtained by actual measurements.

$$P = \frac{\frac{T}{R + Y} \times Y}{A}$$

$$P = \frac{56.40}{.224}$$

$$P = 251 \text{ persons}$$

Carrying Capacity, Land under High Forest

Estimates will be made for (1) pig population at minimum and (2) pig population at maximum.

CARRYING CAPACITY, PIG POPULATION AT MINIMUM

$$(1) \quad P = \frac{\frac{T \text{ class III}}{R + Y} \times Y}{A}$$

$$P = \frac{\frac{351 \times .857 \times .43}{14 + 1} \times 1}{A}$$

a) $A = .142$ acres

$$P = \frac{\frac{129}{15}}{.146}$$

$$P = 60 \text{ persons}$$

b) $A = .194$ persons

$$P = \frac{\frac{129}{15}}{.194}$$

$$P = 44 \text{ persons}$$

CARRYING CAPACITY, PIG POPULATION AT MAXIMUM

$$A = .224 \text{ acres}$$

$$P = \frac{\frac{129}{15}}{.224}$$

$$P = 38 \text{ persons}$$

ESTIMATION OF CARRYING CAPACITY FOR PIGS

As in the case of the human population, estimates will be made separately for (1) areas under cultivation and secondary forest during the fieldwork period, and (2) areas under high forest but which were deemed to be arable. Domestic food requirements only are considered. No attempt was made to estimate the requirements of the animals for grazing land.

Carrying Capacity for Pigs, Land under Cultivation and Secondary Forest

Two methods are used as a check upon each other: (1) the number of pigs that could be supported beyond the maximum of 1962 is computed and added to the maximum, and (2) the number of pigs that could be supported above the minimum is added to the minimum.

COMPUTING CARRYING CAPACITY FOR PIGS, BY INCREMENT TO MAXIMUM POPULATION

The following formula (9) may be used:

$$Pp = Nm + Na \quad (9)$$

Where:

Pp = The carrying capacity for pigs, holding human population constant (204 persons distributed in age-sex categories as in Table 1).

Nm = Maximum herd size, as censused (169 pigs, average size 3.8, or 120 to 150 pounds).

Na = Additional pigs, which may be supported without shortening fallows or bringing virgin forest into production. Na may be solved for by the following equation:

$$Na = \frac{T - Gl}{K} \quad (10)$$

T = Total Class I or Class I-equivalent land which may be put into cultivation annually without shortening fallows or bringing virgin forest into production.

$$T = \frac{\text{T Class I} + (\text{T Class II} \times .926 \times .6)}{R + Y} \times Y.$$

Gl = Total acreage of Class I or Class I-equivalent land put into production when the pig population was at maximum size. Values have been assigned by actual measurements of all gardens in 1962.

K = Amount of land required to feed each 120- to 150-pound pig in excess of those included in the minimum pig population. K has been assigned a value of .15 acres (see Chapter 3, p. 62).

These equivalents may now replace unknowns in formula (9).

$$Pp = Nm + Na$$

$$Pp = Nm + \frac{T - Gl}{K}$$

$$Pp = Nm + \frac{\dfrac{\text{T Class I} + (\text{T Class II} \times .926 \times .6)}{(R + Y)} \times Y - [\text{G Class I} + (\text{G Class II} \times .926)]}{K}$$

$$Pp = 169 + \frac{\dfrac{650 + (352 \times .926)}{(14 + 1)} \times 1 - (27.84 + 17.91)}{.15}$$

$$Pp = 169 + \frac{56.40 - 45.75}{.15}$$

$$Pp = 169 + 71$$

$$Pp = 240 \text{ pigs averaging 120 to 150 pounds}$$

COMPUTING CARRYING CAPACITY FOR PIGS BY INCREMENT TO THE MINIMUM POPULATION

The following formula is used:

(11) $$Pp = \frac{Ns}{2} + Nb$$

Where:

Pp = The carrying capacity for pigs of an average size of 120 to 150 pounds, holding human population constant (204 persons distributed in age-sex categories in Table 1).

Ns = The number of pigs, averaging 60 to 75 pounds comprising the minimum pig herd (Nov. 1963).

2 = A correcting factor, converting average size of animals when pig population is minimum (60 to 75 lbs.) to average size when pig population is maximum (120 to 150 lbs.).

Nb = The number of additional pigs of size 120 to 150 pounds which may be supported without shortening fallows or bringing virgin forest into production. Nb may be solved for by the following equation:

$$Nb = \frac{T - Gs}{K}. \quad (12)$$

Where:

T and K have the same values as in formula (10). Gs represents total acreage of Class I or Class I-equivalent land under production when the pig population was at a minimum. Three values have already been derived for Gs. Calculations will be made using the two extremes: 28.87 and 39.61 acres.

a) Value of Gs = 39.61 acres

$$Pp = \frac{Ns}{2} + \frac{T - Gs}{K}$$

$$Pp = \frac{60}{2} + \frac{56.40 - 39.61}{.15}$$

Pp = 142 pigs averaging 120 to 150 pounds.

b) Value of Gs = 28.87 acres

$$Pp = \frac{60}{2} + \frac{56.40 - 28.87}{.15}$$

Pp = 214 pigs averaging 120 to 150 pounds

Carrying Capacity for Pigs, Land under High Forest

Only formula (11) will be used. Adjustment must be made, however, in the value of Ns. This adjustment is accomplished by solving the following ratio for X.

$$\text{Gl:Gh} = \text{Ns:X} \tag{13}$$

Where:

Gl = The maximum amount of Class I and Class II land that could be put under production at one time without shortening fallows. This has already been determined (see p. 293) to be 56.40 acres.

Gh = The maximum amount of Class III land that could be put under production at one time allowing a 35-year fallow. This has already been determined to be 8.6 acres.

Ns = The number of pigs, averaging 60 to 75 pounds, comprising the minimum pig herd (Nov. 1963). This number, by census, was determined to be 60.

$56.40:8.6 = 60:X$

$X = 9.1$

Values may now be provided for all variables in formula

$$Pp = \frac{Ns}{2} + Nb \tag{11}$$

$$Pp = \frac{9.1}{2} + \frac{8.6}{.15}$$

Pp = 62 pigs averaging 120 to 150 pounds.

BIBLIOGRAPHY

Albanese, Anthony A., and Louise A. Orto, 1964. The proteins and amino acids, *Modern Nutrition in Health and Disease,* 3d ed. Michael G. Wohl and Robert S. Goodhart, eds. Philadelphia, Lee and Febiger.

Allan, W., 1949. *Studies in African Land Usage in Northern Rhodesia.* Rhodes-Livingstone Papers #15. Livingstone, Northern Rhodesia, The Rhodes-Livingstone Institute.

Allee, W. C., et al., 1949. *Principles of Animal Ecology.* Philadelphia, Saunders.

Andrewartha, H. E., 1961. *Introduction to the Study of Animal Populations.* Chicago, University of Chicago.

Axelrod, A. E., 1964. Nutrition in relation to acquired immunity, *Modern Nutrition in Health and Disease,* 3d ed. Michael G. Wohl and Robert S. Goodhart, eds. Philadelphia, Lee and Febiger.

Bateson, Gregory, 1936. *Naven.* Cambridge, Eng., Cambridge University.

Berg, Clarence, 1948. Protein deficiency and its relationship to nutritional anemia, hypoproteinemia, nutritional edema, and resistance to infection, *Proteins and Amino Acids in Nutrition.* Melville Sahyun, ed. New York, Reinhold.

Bierstadt, Robert, 1950. An analysis of social power, *American Sociological Review, 15*:730–38.

Birdsell, J., 1958. On population structure in generalized hunting and collecting populations, *Evolution, 12*:189–205.

Blest, A. D., 1961. The concept of ritualisation, *Current Problems in Animal Behavior.* W. H. Thorpe and O. L. Zangwill, eds. Cambridge, Eng., Cambridge University.

Brookfield, H. C., 1964. The ecology of highland settlement: some suggestions, *American Anthropologist,* vol. 66, no. 4, part 2, pp. 20–39.

Brookfield, H. C., and Paula Brown, 1958. Chimbu land and society, *Oceania, 30*:1–75.

——— 1963. *Struggle for Land.* Melbourne, Australia, Oxford University.

Bulmer, Ralph, 1960. Political aspects of the Moka ceremonial exchange system among the Kyaka people of the western highlands of New Guinea, *Oceania, 31*:1–13.

——— 1965. The Kyaka of the western highlands, *Gods, Ghosts and Men in Melanesia.* P. Lawrence and M. J. Meggitt, eds. Melbourne, Australia, Oxford University.

Bulmer, Susan and Ralph, 1964. The prehistory of the Australian New Guinea highlands, *American Anthropologist,* vol. 66, no. 4, part 2, pp. 39–76.

Bunzel, Ruth, 1938. The economic organization of primitive peoples, *General Anthropology.* Franz Boas, ed. New York, Heath.

Burkill, Isaac Henry, 1935. *A Dictionary of the Economic Products of the Malay Peninsula.* London, published on behalf of the governments of the Straits Settlements and Federated Malay States by the crown agents for the colonies.

Burton, Benjamin T., ed., 1959. *The Heinz Handbook of Nutrition.* New York, McGraw-Hill.

Carneiro, Robert L., 1956. Slash and burn agriculture: a closer look at its implications for settlement patterns, *Selected Papers of the Fifth International Congress of Anthropological and Ethnological Sciences.* A. F. C. Wallace, ed. Philadelphia.

——— 1964. Shifting cultivation among the Amahuaca of eastern Peru, *Volkerkundliche Abhandlungen.* Band I. Niedersachsisches Landesmuseum Hannover Abteilung fur Volkerkunde.

Chappell, John, 1966. Stone ax factories in the highlands of East New Guinea, *Proceedings of the Prehistoric Society, 32:* 96–121.

Chapple, E. D., and C. S. Coon, 1942. *Principles of Anthropology.* New York, Holt.

Clark, William, 1966. From extensive to intensive shifting cultivation: A succession from New Guinea, *Ethnology,* 5:347–59.

Clausen, Hjalmar, and Claude Gerwig, 1958. *Pig Breeding, Recording and Progeny Testing in European Countries.* FAO Agricultural Studies #44. Rome, Food and Agriculture Organization of the United Nations.

Collins, Paul, 1965. Functional analyses in the symposium "Man, culture, and animals," *Man, Culture, and Animals.* Anthony Leeds and Andrew P. Vayda, eds. Washington, D. C., American Association for the Advancement of Science 78.

Conklin, H. C., 1957. *Hanunoo Agriculture in the Philippines.* FAO Forestry Development Paper #12. Rome, Food and Agriculture Organization of the United Nations.

——— 1961. The study of shifting cultivation, *Current Anthropology, 2:*27–64.

Cook, S. F., 1946. Human sacrifice and warfare as factors in the demography of pre-colonial Mexico, *Human Biology, 18:*81–100.

Duncan, Otis Dudley, 1961. From social system to ecosystem, *Sociological Inquiry, 31:*140–49.

Durkheim, Emile, 1912. *Les formes élémentaires de la vie religieuse.* Paris, Alcan.

——— 1938. *The Rules of Sociological Method.* 8th ed. Sarah A. Solovay and John H. Mueller, trans. George E. G. Catlin, ed. Glencoe, N.Y., Free Press.
——— 1961. *The Elementary Forms of the Religious Life.* John Ward Swain, trans. New York, Collier.

Elman, Robert, 1951. *Surgical Care. A Practical Physiologic Guide.* New York, Appleton-Century-Crofts.
Etkin, Wm., 1963. Theories of socialization and communication, *Social Behavior and Evolution among the Vertebrates.* Wm. Etkin, ed. Chicago, University of Chicago.

Fei, Hsiao-Kung, 1945. *Earthbound China.* Chicago, University of Chicago.
Firth, Raymond, 1929. *Primitive Economics of the New Zealand Maori.* London, Routledge.
——— 1950. *Primitive Polynesian Economy.* New York, Humanities Press.
Food and Agriculture Organization of the United Nations, 1950. *Calorie Requirements. Report of the Committee on Calorie Requirements.* FAO Nutritional Studies #5. Rome, FAO.
——— 1955. *Protein Requirements.* FAO Nutritional Studies #16 Rome, FAO.
——— 1957. *Calorie Requirements. Report of the Second Committee on Calorie Requirements.* FAO Nutritional Studies #15. Rome, FAO.
——— 1964. *Protein: At the Heart of the World Food Problem.* World Food Problems #5. Rome, FAO.
Frake, Charles O., 1962. Cultural ecology and ethnography, *American Anthropologist, 64:*53–59.
Frank, Lawrence K., 1966. Tactile communication, *Culture and Communication.* Alfred G. Smith, ed. New York, Holt, Rinehart, Winston. Reprinted from *Genetic Psychology Monographs,* 56:209–55.
Freeman, J. D., 1955. *Iban Agriculture. A Report on the Shifting Cultivation of Hill Rice by the Iban of Sarawak.* London, Her Majesty's Stationery Office.
Freud, Sigmund, 1907. Obsessive actions and religious practices, Z. *Religionspsychol.,* vol. I, no. 1, pp. 4–12, trans. in *The Standard Edition of the Complete Psychological Works of Sigmund Freud.* IX. James Strachey, gen. ed. London, Hogarth, 1959.

Glasse, Robert, 1959. Revenge and redress among the Huli: A preliminary account, *Mankind,* 5:273–89.
Gluckman, Max, 1954. *Rituals of Rebellion in Southeast Africa.* The Frazer lecture, 1952. Manchester, Eng., Manchester University.
——— 1962. Les rites de passage. *The Ritual of Social Relations.* Max Gluckman, ed. Manchester, Eng., Manchester University.
Goffman, Erving, 1956. The nature of deference and demeanor, *American Anthropologist, 58:*473–503.
Goldman, Stanford, 1960. Further consideration of cybernetic aspects of ho-

meostasis, *Self-Organizing Systems*. M. C. Yovits and Scott Cameron, eds. New York, Pergamon Press.

Goodenough, Ward, 1955. A problem in Malayo-Polynesian social organization, *American Anthropologist*, 57:71–83.

Hagen, Everett E., 1962. *On the Theory of Social Change*. Homewood, Ill., Dorsey.

Hamilton, L., 1955. Indigenous versus introduced vegetables in the village dietary, *The Papua and New Guinea Agricultural Journal*, 10:54–57.

Harris, Marvin, 1965. The myth of the sacred cow, *Man, Culture, and Animals*. Anthony Leeds and Andrew P. Vayda, eds. Washington, D.C., American Association for the Advancement of Science.

——— n.d. Cultural energy. Unpublished paper.

Haswell, M., *Economics of Agriculture in a Savannah Village*. Colonial Research Studies #8. London, H.M.S.O.

Hatfield, Charlotte, 1954. *Food Composition Tables—Minerals and Vitamins—for International Use*. FAO Nutritional Studies #11. Rome, FAO.

Hawley, Amos, 1944. Ecology and human ecology, *Social Forces*, 22:398–405.

Helm, June, 1962. The ecological approach in anthropology, *The American Journal of Sociology*, 67:630–39.

Hempel, Carl, 1959. The logic of functional analysis, *Symposium on Sociological Theory*. L. Gross, ed. Evanston, Ill., Row, Peterson.

Hendrix, Gertrude, Jacques D. Van Vlack, and William Mitchell, 1966. *Equine-Human Linked Behavior in the Post-natal and Subsequent Care of Highly-Bred Horses*. Paper and film presented to the annual meeting of the American Association for the Advancement of Science, Dec. 26, 1966.

Hinde, R. A., and N. Tinbergen, 1958. The comparative study of species specific behavior, *Behavior and Evolution*. A. Roe and G. G. Simpson, eds. New Haven, Yale University.

Hipsley, E. H., and F. W. Clements, 1947. *Report of the New Guinea Nutrition Survey Expedition*. Canberra, Australia, Department of External Territories.

Hipsley, E. H., and Nancy Kirk, 1965. *Studies of Dietary Intake and the Expenditure of Energy by New Guineans*. South Pacific Commission Technical Paper #147. Noumea, New Caledonia.

Hitchcock, Nancy, 1962. Composition of New Guinea Foodstuffs. Unpublished.

Homans, George C., 1941. Anxiety and ritual: The theories of Malinowski and Radcliffe-Brown, *American Anthropologist*, 43:164–72.

Houssay, Bernardo Alberto, et al., 1955. *Human Physiology*, 2d ed. Juan T. Lewis and Olive T. Lewis, trans. New York, McGraw-Hill.

Izikowitz, Karl Gustav, 1951. *Lamet, Hill Peasants in French Indochina*. Göteborg, Sweden, Etnografiska Museet.

Jay, Phyllis, 1963. The Indian Langur Monkey, *Primate Social Behavior*. Charles H. Southwick, ed. Princeton, Van Nostrand.

Kluckhohn, Clyde, 1944. *Navaho Witchcraft*. Papers of the Peabody Museum of American Archaeology and Ethnology 22, #2. Cambridge, Mass., Harvard University.

Kroeber, A., 1939. *Cultural and Natural Areas of North America*. Berkeley, University of California.

Langley, Doreen, 1947. Part 4, *Report of the New Guinea Nutrition Survey Expedition*. E. H. Hipsley and F. W. Clements, eds. Canberra, Australia, Department of External Territories.

Large, Alfred, and Charles G. Johnson, 1948. Proteins as related to burns, *Proteins and Amino Acids in Nutrition*. Melville Sahyun, ed. New York, Reinhold.

Lawrence, Peter, 1964. *Road Belong Cargo*. Manchester, Eng., Manchester University.

Leach, E. R., 1954. *Political Systems of Highland Burma*. Boston, Beacon.

Linton, Robert, 1955. *The Tree of Culture*. New York, Knopf.

Livingstone, Frank B., 1958. Anthropological implications of sickle cell gene distribution in West Africa, *American Anthropologist*, 60:533–62.

Loffler, Lorenz G., 1960. Bodenbedarf und Ertragsfaktor in Brandrodungsbau, *Tribus*, 9:39–43.

Lund, Charles C., and Stanley M. Levenson, 1948. Protein nutrition in surgical patients, *Proteins and Amino Acids in Nutrition*. Melville Sahyun, ed. New York, Reinhold.

Luzbetak, Louis J., 1954. The socio-religious significance of a New Guinea pig festival, parts I and II, *Anthropological Quarterly*, 2:59–80, 102–28.

Malinowski, Bronislau, 1948. *Magic Science and Religion and Other Essays*. Boston, Beacon.

Massal, E., and Jacques Barrau, 1956. *Food Plants of the South Sea Islands*. South Pacific Commission Technical Paper #94. Noumea, New Caledonia.

Meggitt, M. J., 1958. The Enga of the New Guinea highlands: some preliminary observations, *Oceania*, 28:253–330.

——— 1962. Growth and decline of agnatic descent groups among the Mae Enga of the New Guinea highlands, *Ethnology*, 1:158–67.

——— 1965. *The Lineage System of the Mae Enga of New Guinea*. Edinburgh and London, Oliver and Boyd.

Merton, Robert K., 1949. *Social Theory and Social Structure*. Glencoe, N.Y., Free Press.

Moore, Francis D., 1959. *Metabolic Care of the Surgical Patient*. Philadelphia and London, Saunders.

Moore, Oman Khayyam, 1957. Divination—A new perspective, *American Anthropologist*, 59:64–74.

Miller, Robert J., 1964. Cultures as religious structures, *Symposium on New Approaches to the Study of Religion*. Proceedings of the 1964 annual spring meeting of The American Ethnological Society. June Helm, ed. Seattle, University of Washington.

National Academy of Sciences, 1963. *Evaluation of Protein Quality*. National Research Council Publication 1100. Washington, D.C., NAS-NRC.

Newman, Philip, 1964. Religious belief and ritual in a New Guinea society, *American Anthropologist*, vol. 66, no. 4, part 2, pp. 257–72.

Newton, Kenneth, 1960. Shifting cultivation and crop rotations in the tropics, *Papua and New Guinea Agricultural Journal*, 13:81–118.

Odum, Eugene P., 1959. *Fundamentals of Ecology*, 2d ed. Philadeilphia, Saunders.

Oliver, Douglas, 1955. *A Solomon Island Society; Kinship and Leadership among the Siuai of Bougainville*. Cambridge, Mass., Harvard University.

Oomen, H. A. P. C., 1961. The nutrition situation in western New Guinea, *Journal of Tropical Geography and Medicine*, 13:321–35.

——— 1961. The Papuan child as a survivor, *The Journal of Tropical Pediatrics and African Child Health*, 6:103–21.

Oosterwal, G., 1961. *People of the Tor: A Cultural-Anthropological Study on the Tribes of the Tor Territory*. Assen, Netherlands, Koninklijke Van Gorcum.

Osmund, A., and W. Wilson, 1961. *Tables of Composition of Australian Foods*. Canberra, Australia, Commonwealth Department of Health.

Peters, F. E., 1958. *The Chemical Composition of South Pacific Foods*. South Pacific Commission Technical Paper #115. Noumea, New Caledonia.

Powers, W. T., R. K. Clark, and R. L. McFarland, 1960. A general feedback theory of human behavior, *Perceptual and Motor Skills*, 11:71–88. Reprinted in *Communication and Culture*. Alfred G. Smith, ed. New York, Holt, Rinehart, Winston, 1966.

Radcliffe-Brown, A. R., 1952. Religion and society, *Structure and Function in Primitive Society*. Glencoe, N.Y., Free Press.

Rappaport, Roy A., 1963. Aspects of man's influence upon island ecosystems: alteration and control, *Man's Place in the Island Ecosystem*. F. R. Fosberg, ed. Honolulu, Bishop Museum.

——— 1966. Ritual in the ecology of a New Guinea people. Columbia University, doctoral dissertation.

——— 1967. Ritual regulation of environmental relations among a New Guinea people, *Ethnology*, 6:17–30.

Read, Kenneth, 1952. Nama cult of the central highlands, New Guinea, *Oceania*, 33:1–25.

Reay, Marie, 1959. *The Kuma: Freedom and Conformity in the New Guinea Highlands*. Melbourne, Australia, Melbourne University Press for the Australian National University.

Reik, Theodor, 1962. *Ritual: Four Psychoanalytic Studies*. Douglas Bryan, trans. New York, Grove Press.

Richards, P. W., 1964. *The Tropical Rain Forest. An Ecological Study*. Cambridge, Eng., The University Press.

Robbins, Ross G., 1961. The vegetation of New Guinea, *Australian Territories,* vol. 1, no. 6, pp. 1–12.

——— 1963. The anthropogenic grasslands of New Guinea, *Proceedings of the* UNESCO *Symposium on Humid Tropic Vegetation. Goroka, 1960.* Canberra, Australia, Government Printer.

Ryan, D'arcy, 1959. Clan formation in the Mendi Valley, *Oceania, 29:*257–90.

Sahlins, Marshall David, 1958. *Social Stratification in Polynesia.* Seattle, University of Washington.

——— 1963. Poor man, rich man, big-man, chief: Political types in Melanesia and Polynesia, *Comparative Studies in Society and History, 5:*285–303.

——— 1965. Exchange-value and the diplomacy of primitive trade, *Essays in Economic Anthropology.* Proceedings of the American Ethnological Society. June Helm, ed. Seattle, University of Washington.

Salisbury, Richard F., 1962. Ceremonial economics and political equilibrium, *Actes du VIe Congres International des Science Anthropologiques et Ethnologiques.* I. Paris.

Schwartz, Theodore, 1962. Systems of areal integration: some considerations based on the Admiralty Islands of northern Melanesia, *Anthropological Forum, 1:*56–97.

Simpson, George Gaylord, 1962. Comments on cultural evolution, *Evolution and Man's Progress.* H. Hoagland and R. W. Burhoe, eds. New York, Columbia University.

Stott, D. H., 1962. Cultural and natural checks on population growth, *Culture and the Evolution of Man.* M. F. Ashley Montagu, ed. New York, Oxford University.

Street, John, 1965. Evaluation of the concept of carrying capacity. Unpublished paper, delivered at meeting of Association of Pacific Coast Geographers, Portland, Ore., June 18, 1965.

Tinbergen, N., 1952. The evolution of animal communication—a critical examination of methods, *Symposium of the Zoological Society of London, 8:*1–6.

——— 1963. The evolution of signalling devices, *Social Behavior and Organization among the Vertebrates.* Wm. Etkin, ed. Chicago, University of Chicago.

Vayda, Andrew P., and E. A. Cook, 1964. Structural variability in the Bismarck Mountain cultures of New Guinea: a preliminary report, *Transactions of the New York Academy of Sciences.* ser. II, vol. 26, no. 7, pp. 798–803.

Vayda, Andrew P., Anthony Leeds, and David Smith, 1961. The place of pigs in Melanesian subsistance, *Proceedings of The American Ethnological Society.* Viola E. Garfield, ed. Seattle, University of Washington.

Vayda, Andrew P., and Roy A. Rappaport, 1963. Island cultures, *Man's Place in the Island Ecosystem.* F. R. Fosberg, ed. Honolulu, Bishop Museum.

——— in press. Ecology, cultural and non-cultural, *Introduction to Cultural*

Anthropology: Essays in the Scope and Methods of the Science of Man. James A. Clifton, ed. Boston, Houghton Mifflin.

Venkatachalam, P. S., 1962. *A Study of the Diet, Nutrition and Health of the People of the Chimbu Area.* Monograph #4. Territory of Papua and New Guinea, Department of Public Health.

Waddington, C. H., 1960. Panel three: Man as an organism, *Evolution after Darwin,* vol. III, *Issues in Evolution.* Sol Tax, ed. Chicago, University of Chicago, pp. 171–73.

Wallace, Anthony F. C., 1966. *Religion: An Anthropological View.* New York, Random House.

White, Leslie, 1949. *The Science of Culture.* New York, Farrar Straus.

Wolf, Eric R., 1955. Types of Latin American peasantry: A preliminary discussion, *American Anthropologist,* 57:452–71.

Wurm, Stephan, 1964. Australian New Guinea Highlands languages and the distribution of their typological features, *American Anthropologist,* vol. 66, no. 4, part 2, pp. 77–97.

Wynne-Edwards, V. C., 1962. *Animal Dispersion in Relation to Social Behavior.* Edinburgh and London, Oliver and Boyd.

——— 1965. Self regulating system in populations of animals, *Science, 147:* 1543–47.

Zintel, Harold A., 1964. Nutrition in the care of the surgical patient, *Modern Nutrition in Health and Disease,* 3d ed. Michael G. Wohl and Robert S. Goodhart, eds. Philadelphia, Lee and Febiger.

Zucker, Theodore F., and Lois M. Zucker, 1964. Nutrition and natural resistance to infection, *Modern Nutrition in Health and Disease,* 3d ed. Michael G. Wohl and Robert S. Goodhart, eds. Philadelphia, Lee and Febiger.

INDEX